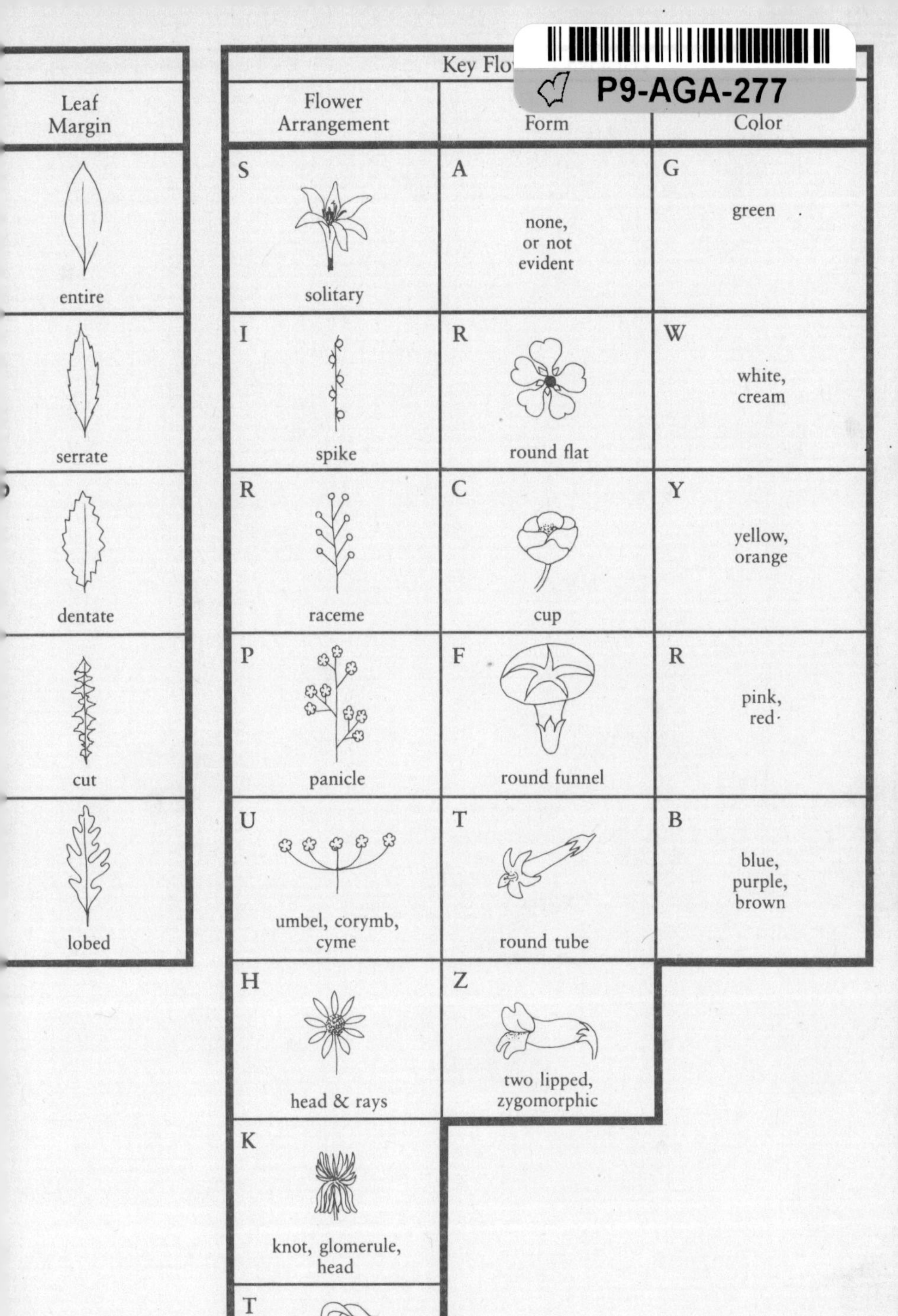

Leaf Margin
entire
serrate
dentate
cut
lobed
Key Flo
Flower Arrangement
Form
Color
S
solitary
A
none, or not evident
G
green
I
spike
R
round flat
W
white, cream
R
raceme
C
cup
Y
yellow, orange
P
panicle
F
round funnel
R
pink, red
U
umbel, corymb, cyme
T
round tube
B
blue, purple, brown
H
head & rays
Z
two lipped, zygomorphic
K
knot, glomerule, head
T
spathe
P9-AGA-277

Florida Wild Flowers and Roadside Plants

Published with the Sponsorship of

The Florida Federation of Garden Clubs, Inc. *and* The Florida Audubon Society, Inc.

Laurel Hill Press
Chapel Hill
1982

Florida Wild Flowers

and Roadside Plants

C. Ritchie Bell
Bryan J. Taylor

LIBRARY OF CONGRESS CATALOGING IN PUBLICATION DATA

Bell, C. Ritchie.
Florida wild flowers and roadside plants.

Includes index.
1. Wild flowers—Florida—Identification.
2. Roadside flora—Florida—Identification.
I. Taylor, Bryan J., 1942– . II. Title.
QK154.B44 1982 582.13'09759 82-16234
ISBN 0-9608688-0-1

First Printing 1982

Book designed and produced by
JOYCE KACHERGIS BOOK DESIGN AND PRODUCTION
Bynum, N.C.

Acknowledgments

Few books are completely original or are ever produced by the authors alone. This is especially true for a book of this scope which, though new, builds not only on the field experience of the authors but also on several centuries' accumulation of general botanical knowledge and on the present helpful assistance, and generous contributions, of many people during various stages of photography, writing and production of *Florida Wild Flowers*. The authors acknowledge, with gratitude, their debt to all who have assisted this project, either directly or indirectly, over the years. It is our hope that our efforts prove worthy of this generous help. Also, it is especially appropriate to acknowledge here the early support of this project by Mrs. Bernese B. Davis, Mrs. Hilda Kressman, Mrs. Helen Hood, and the late Mrs. Velma F. Gwinn of the Florida Federation of Garden Clubs, Mr. William Partington of the Environmental Information Center, and Mr. Dade Thornton and Mr. Hal Scott of the Florida Audubon Society.

A number of friends and fellow naturalists contributed much to the productivity and enjoyment of the photographic field work. It is a pleasure to acknowledge with thanks the warm personal hospitality, and the often extensive field assistance of Major James A. Stevenson, Mr. Kenneth C. Alvarez, Mrs. Jeane Parks, and Mr. Larry Sullivan of the Florida Division of Parks and Recreation; Mr. Oren "Sonny" Bass and Dr. Ingrid C. Olmsted of the Everglades National Park; Paul and Lee Mueller of Naples; and Mrs. Mary Steffee Degtoff of Kissimmee. Their help was most appreciated.

A significant portion of the floral beauty and botanical diversity included in this book results from the wonderful cooperation of a small number of excellent photographers who have generously made some of their choice slides of Florida plants available to the authors. These special slides enabled us to increase the total coverage of the book to include a number of critical botanical groups or geographic areas of Florida that would not have been possible otherwise. We offer our profound thanks to each of these individuals, whose names are given in the following section and whose initials occur by their photographs throughout the book, and also to Ms. Dorothy Wilbur, of the North

Carolina Botanical Garden, from whose skilled pen came the line drawings used to illustrate some of the plant terms.

The technical collaboration of Dr. Anne H. Lindsey at every phase of the project, including field assistance, library and herbarium searches, and computerization of the species character codes for the key construction has been a critical element in the production of this book and deserves the very special thanks hereby tendered.

No contribution to a book of this type is more important, or more appreciated, than the careful critical review of the manuscript for technical scientific accuracy. Such detailed reviews were kindly provided by Mr. Steven W. Leonard and Mr. Allen G. Shuey of Bradenton, Florida; many entries have benefited from their extensive personal knowledge of the Florida flora and from their interesting comments and appropriate questions on the early pages of manuscript. In addition, Dr. Jimmy Massey, Dr. Albert E. Radford, Dr. Robert Peet, Dr. Rogers McVaugh, Dr. Michael I. Cousens, Dr. John Poponoe, and Dr. Daniel Ward all provided assistance from time to time with the verification of factual information concerning particular plants or the identification of questionable specimens. The many helpful comments of each of these professional colleagues is most gratefully acknowledged. With such erudite assistance, any errors as may remain in the text are clearly the responsibility of the authors, and we would welcome the opportunity to correct them.

So long as this book brings enjoyment or provides information to anyone interested in Florida's varied flora, both the authors and the users will owe continuing thanks to the officers and members of the Florida Federation of Garden Clubs and the Florida Audubon Society for their foresight, interest, and efforts that made this publication possible.

C.R.B.
B.J.T.
March 1982

Contents

Introduction

Nearly five centuries ago, in the spring of 1513, the great variety, abundance, and beauty of the many native flowers of this new land so impressed the explorer Ponce de Leon that he called it Florida—the land of flowers. The name still applies today. Indeed, because of the many weeds and exotic new horticultural plants either intentionally or unintentionally introduced into the state, there are doubtless more different kinds of plants growing wild in Florida today than when it was first explored by Europeans. There is no actual enumeration of all of the kinds of plants in the entire state, but the total number of ferns, conifers, and flowering species has been currently estimated at about 3,500. This number includes all of the trees, shrubs, vines, ferns, grasses, sedges, and herbs to be found growing in the woods, swamps, groves or fields, and along the roadsides of Florida.

Most of these 3,500 species of plants are native, but many are introduced. Some have flowers that are quite colorful and showy; others may have small flowers that are seldom seen, or seldom noticed. Many species are both widespread and common, but several are endangered, threatened, or rare and are now protected by law. Different plants are found in different regions of the state and in different habitats, and in Florida's gentle climate, there is no season without some plants in flower.

Five hundred of these species, representing 137 plant families are illustrated in this work. An additional 215 species are treated in the text; thus about one fifth of Florida's total flora is included. If consideration is given to the many related plants that, though not illustrated or mentioned, can be adequately identified by association (see examples under item 5 of format), it is likely that this book would aid in the identification to genus of over half of the more showy and frequent flowers found in Florida.

Although variation in climate, from warm temperate to subtropical or tropical, is an important aspect of the great diversity of the Florida flora, geology and geography also play a role in plant distribution. The small but interesting series of species that are truly endemic to Florida are the most unique elements of Florida's varied flora. Perhaps a hundred or so species of more northern hardwood forests follow the drainage patterns and clay-based soils from the southern Appalachians into the wooded hills along the upper reaches of the Apalachicola River. Likewise, many tropical plants have migrated from island to island to reach their northern limit in the Florida Keys or the Everglades, and plant migrations along the Gulf Coast (often greatly assisted by man's activity!) continue to bring new plants into the Panhandle from as far away as Texas or even Mexico. In relation to its size, Florida may well have a more varied flora than any similar area of North America above the Rio Grande. The diversity of Florida's plants and plant communities, with their associated animal dependents, provides an almost infinite array of interesting subject matter for all who enjoy any aspect of the fascinating field of natural history, whether at the level of a recreational walk along the beach or a woodland trail, an informal nature study course, or highly structured scientific research.

DESCRIPTIVE FORMAT

Although a picture is indeed worth a thousand words, a few appropriate words are often very helpful in plant identification. The text for each species illustrated in this book is necessarily brief but provides, in combination with the distribution map, blooming season symbol, and character code information, a relatively complete description useful to readers with various levels of botanical knowledge and interest. With this information, most of the plants represented in the book can be identified with a higher degree of certainty than is possible through the picture alone. The sixteen elements that make up a typical entry for each photograph are listed in sequence below. Comments on the use or application of each information element and definitions, as may be appropriate, are given in subsequent paragraphs. Please note that elements 4–9, which are treated within the text for each species, may not always be in the same sequence, but may vary somewhat depending on context.

SUMMARY OF DESCRIPTIVE FORMAT ELEMENTS

1. Species number and common name
2. Asterisk denotes protected status (State or Federal)
3. Scientific name (genus, species, authority)
4. Measurement to give scale to photograph
5. Notes on related species, if included
6. Frequency, where appropriate
7. Habitat, native or naturalized status
8. General distribution in Florida (also see map for quick reference)
9. General distribution outside Florida
10. Photographers initials
11. Nine-part key character code
12. Reference for more information
13. Plant family
14. Three centimeter reference scale
15. Map for Florida distribution
16. Season of bloom indicator

1–2. Species number, common name, and legal status

The arrangement of the 500 plant species illustrated follows the general botanical classification system for the major plant families which place the non-flowering ferns, and the pines (which are gymnosperms), before the flowering plants or angiosperms. Within the angiosperms, the grasses, sedges, lilies, orchids, and other families of monocotyledons are grouped together apart from the numerous families of dicotyledons such as the water lilies, roses, mints, milkweeds, and asters. For convenience, the genera are listed alphabetically within each family, as are the species under each genus. Within the above system the illustrated species are numbered sequentially from 1 to 500. These "FWF Species Numbers" greatly facilitate indexing and rapid reference, and their use saves considerable space in keys and checklists.

Many of our native plants have one or more common names that can be very colorful and descriptive and are sometimes easier to remember than a scientific name, thus the first name given for each entry is a common name. However, since there can be several common names for a single species, such names are easily misapplied, can often be confusing, and may actually give inaccurate information about a plant. For example, the weedy shrub, *Callicarpa americana*, often called 'Beauty Berry' or 'French Mulberry' is neither French nor is it a Mulberry; it is a North American member of the Verbena family. Often the common name is merely a translation of a Latin scientific name, such as Catesby's Lily for *Lilium catesbaei*; or the Latin genus name may also function as an accepted common name, or part of a common name, for several different species in the same genus, as in the case of the various species of *Iris, Magnolia, Aster, Rhododendron, Liatris,* and *Chrysanthemum,* all of which are used as common plant names with equal ease and familiarity by botanists and nonbotanists alike.

An asterisk (*) after the common name indicates that the species is rare, threatened or endangered and is protected by either federal or state law and should be left undisturbed in its native habitat to help conserve the species and for the enjoyment of others.

3. Scientific Name

The Latin or Latinized scientific name for each species, or kind, of plant is given under the common name. The scientific (or species) name always consists of two words: a genus or generic name (e.g. *Magnolia*) and a species epithet (e.g., *Magnolia virginiana*). As a bibliographic reference, this two-part, or binomial, name is followed by the name of the botanist who first described and named the plant. Since the Swedish botanist Linnaeus first published the name used in our example above (in 1753), the complete citation of the scientific name, including the author or authority, would be *Magnolia virginiana* Linnaeus; the plant is also known by its common name, "Sweet Bay."

Unlike common names, the scientific name must be unique, and the naming of plants is carefully prescribed by a set of international rules called the Code of Botanical Nomenclature. Although many plants may have the same specific epithet, or species name, they must have different generic names since each species must have a distinctive scientific name that is not duplicated anywhere else in the plant kingdom. For example, five plants in this book with the specific epithet florida, or "flowering," include: *Cornus florida* (Flowering Dogwood), *Calycanthus floridus* (Sweet Shrub), *Illicium floridanum* (Star Anise), *Viola floridana* (Florida Violet), and *Stachys floridana* (Hedge Nettle), all of which are not only in different genera, but also happen to be in different families.

Because of occasional historical error, or because of different interpretations of relationship and species delimitation by different botanists, it is possible for two scientific names to be applied to a single species to accommodate the two concepts of relationship. Indeed, the scientific names in botanical references that include Florida often vary from one author to the next because parts of the flora have not yet been studied extensively and valid uncertainty exists as to the proper classification of some species. Though perhaps unfortunate, such synonymous names, as they are called, are a part of the continuing process of plant classification. Even with such synonyms, however, scientific names are far more stable and accurate than common names and should be used as frequently as possible for effective communication. In order to avoid taxonomic confusion, we have, with few exceptions, followed the scientific names in the 1980 "*Synonymized Checklist of the Vascular Flora of the United States, Canada and Greenland*" by John T. Kartesz and Rosemarie Kartesz. In some cases, therefore, the scientific name we use may be different from the synonymous name referenced in other treatments.

4. Scale or Plant Size

To provide an approximate scale for each photograph, each descriptive entry usually contains a specific size reference to a leaf, flower or other plant part in the picture. These measurements are always helpful, and sometimes critical, in identification because similar and closely related species may differ only in some aspect of size. As is usual now in botanical references, all measurements are in metric units. However, to help picture the actual size and to assist with the conversion from English to metric, a three centimeter reference scale (about 1¼ inches) is marked off adjacent to the small distribution map for each entry and a 15 centimeter scale (or 1.5 decimeters, which is essentially 6 inches) is printed with the plant character chart inside the back cover.

5. Notes on Related Species

Species closely related to those illustrated in this book often differ from them by size, relatively minor morphological features, or by obvious differences in habitat. As space permits, one or more of these related species may be mentioned, along with their distinguishing characteristics, as a parenthetical entry in the text. There are 215 such entries which should proportionately extend the usefulness and interest of this book. Indeed, in some instances, the species of a genus are so similar in general aspect that once any one of the species of a genus can be recognized all can be recognized and the same generic or common name can be correctly applied to each of them. For example, there are 13 different species of Violets in Florida and, although the flowers vary a bit in color and size, they all *look* like Violets and can safely be placed in the genus *Viola*. Other examples of such botanical leverage can be found in *Aster* (Asters) with 12 or more species, *Xyris* (Yellow-eyed Grass) with 15 species, *Hypericum* (St. John's Wort) with 16 species, *Solidago* (Goldenrod), and several others. Of course Florida also has many native and naturalized plants represented in the state by only a single species that must be learned individually. The key character chart inside the back cover will be of assistance.

6. Frequency

The general terms denoting frequency (common, frequent, infrequent, rare) are all very subjective and often difficult to apply to a widespread species. A plant common over much of its range is obviously less frequent or rare at the edges of its range. Some plants are indeed quite common in the appropriate habitat, whereas others are quite rare. Thus, where the information is available, some comment as to frequency is stated. Many of our truly rare plants are now protected by either state or federal law, or both, and, as noted previously, such protected plants are marked with an asterisk (*) after the common name and should not be molested in any way.

7. Habitat and Native Status

As indicated previously, the varied flora of Florida is composed of the truly native vegetation, the many introduced plants that have become completely naturalized, and a number of horticultural or agricultural plants that occasionally escape or that may persist for a time around old homesites or abandoned gardens. Because of these varied origins, an attempt has been made to indicate, in each description, whether the plant is native, introduced and naturalized, or an occasional escape, or remnant, from cultivation. If there is no comment about the status of a particular species, it is presumed to be native.

Quite often the habitat in which a particular kind of plant grows is so specific that habitat information becomes a valuable, and occasionally even critical, component of species identification. Various combinations of different environmental conditions such as soil texture and chemical composition, rainfall, and soil moisture content, amount of shade or available light, and daily and seasonal temperature fluctuations, all interact to provide a mosaic of habitats each with its characteristic association of specific plant species. Indeed, such dominant plant associations are often used to identify many of these specialized habitats: hardwood forest, Turkey Oak-Wire Grass sandhill, Mangrove swamp, Cypress swamp, Pine flatwoods, and so on. Each area of the country also has other special habitats related to local geography and geological history. In Florida this would include such habitats as salt water marshes, coastal strands, shell mounds, sinkholes, sinkhole ponds, glades, and bayheads. In addition, man's activities provide another series of habitats, often open and with disturbed soil, such as pastures, fields, groves, roadsides, ditches, sloughs, canals, and spoil banks.

As might be expected, our rarest plants often have the most specialized habitat requirements, and most of our worst weeds seem to be able to grow equally well in a wide variety of habitats. In either case, habitat information is often helpful in species identification and is given for each entry.

There is considerable diversity of opinion among professional ecologists as to the most appropriate classification of the many different habitats and plant associations to be found even in such a relatively small area of the earth's surface as a single state. However, there seems to be at least a general consensus regarding the characteristics and terminology for 17 of Florida's major habitats most frequently encountered. Following the preface of volume 5, on plants, of the *Rare and Endangered Biota of Florida*, eight relatively distinct terrestrial or upland habitats and six wetland habitats are recognized herein. Both the upland and wetland categories are subdivided into open and forest habitats to provide a classification with four major terrestrial categories as described below:

I. UPLAND
  A. Open
    1. Coastal Strand: the open sandy dunes or low bars just above the beaches; usually occupied by succulent, salt-tolerant herbs, vines, and shrubs or small trees.
    2. Dry Prairies: open, dry, or only rarely flooded, grassy flatlands. Most such areas are now improved pastures.
    3. Savanna: generally open, moist (high water table), acid, coastal plain flatland with numerous species of perennial herbs and few, if any, widely scattered trees (also see nos. 4 and 11 below).
  B. Forest
    Pine
    4. Pine Flatwoods: rather open pine woodlands, often with an understory of Palmetto (*Sabal minor*), Gallberry (*Ilex glabra*), and other shrubs with various herbaceous plants. If the shrub layer is sparse and the understory mostly herbs, the area may be called a pine savanna, or if the trees are absent, just savanna (see no. 3 above).
    5. Sand Scrub: thickets or forests of Sand Pine (*Pinus clausa*) on dry sand ridges or old dunes.
    6. Turkey Oak Sandhill: usually an association of Long-leaf Pine (*Pinus palustris*), Turkey Oak (*Quercus laevis*), and Wire Grass (*Aristida stricta*) on hilly, well-drained sands. Much of the original sandhill area of central Florida is now planted in citrus.

Hardwood

7. Mixed Pine-Hardwood: a mixture of pines and deciduous hardwoods; usually found on the clay soils of the Panhandle.
8. Hardwood Hammocks: hardwood forests on islands of slightly higher, and drier, soil in pine flatwoods, prairies, or glades. Bayheads (mostly northern Florida) are stands of swamp hardwoods at the base of a seepage or drainage area. Pocosins are shrub islands, often of Red Bay (*Persea borbonia*) and Sweet Bay (*Magnolia virginiana*).
9. Tropical Hammocks: forests of tropical hardwoods and often Bays or other broadleaf evergreens on islands of drier soil in the glades of south Florida.

II. WETLAND

A. Open

10. Coastal Marshes: saline to brackish tidal water marshes of grasses and sedges.
11. Wet Prairie: periodically flooded and usually moist, open grasslands (also see Savanna, nos. 3 and 4 above).
12. Fresh Water Marshes: low, flat, normally flooded areas with grasses, sedges, other herbs, and a few shrubs.
13. Cypress Scrub: stunted Cypress (*Taxodium distichum*) on flat, flooded, rocky marl soils; where there is better soil, the trees are taller and appear as little hills, or Cypress domes.

B. Forest

14. Cypress Swamp: low areas, often along rivers or lakes, with good soil and full-sized Cypress trees; the mature trees of most such areas have been cut for timber and only smaller second-growth trees remain.
15. Hardwood Swamps or Swamp Forest: low, usually flooded areas along rivers or lakes with forests of Bay (*Magnolia*), Gum (*Nyssa*), and often some Cypress.
16. Mangrove Swamp: areas of Mangrove (*Rhizophora mangle*) thickets in shallow coastal marshes.

III. AQUATIC

17. Open, freshwater springs, lakes, ponds, rivers and streams.

8. Geographic Distribution

Florida clearly has four major phytogeographic regions that can be identified primarily on the basis of climate and soils. These regions are West Florida, North Florida, Central Florida, and South Florida, including the Keys. Although the regions themselves are relatively distinct, there is considerable variation in just where the boundaries between them are drawn. For our purposes it seemed practical to consider the area of South Florida as that defined by Long and Lakela in their *Flora of Tropical Florida* and then define the boundaries of the other regions as shown in Figure 1. In the text, south and central Florida are referred to collectively as peninsular Florida, north and west Florida may be referred to collectively as northern Florida, and west Florida is sometimes called the Panhandle. The Keys are considered part of south Florida but, if a plant species is restricted to the Keys, this is noted.

FIGURE 1. Map of Florida showing the counties and the four major floristic regions. A dot (see small map) indicates the presence of a species in one or more counties of that particular region.

Interestingly, as shown in the tabulation below, each of the four sections of Florida have roughly the same total number of species illustrated in this book. However, the sections vary in the number of endemics and in the number of species found in "one area only." In the latter case, these apparently rare Florida plants may represent species with wide ranges outside of Florida that reach the limit of their respective ranges here as floristic components in one or two counties.

Section of Florida	*Total Number of Treated Species*	*Number of Species Only in This Section*
West	392	27
North	374	4
Central	389	2
South	336	36
All Four	234	—

9. Distribution Outside Florida

With the exception of those few plants that are truly endemic to Florida, and found nowhere else on earth, the diverse Florida flora shows, as expected, strong affinites to the floras of the adjacent states to the north and west, and to the tropical islands to the east and south. Thus to give some idea of the general range, or possible origin, of a particular Florida species, the general distribution of the plant outside of Florida is given. Many species in Florida are found westward along the Gulf coastal plain into Texas, north along the Atlantic coastal plain into the Carolinas and beyond, or they are known to occur also in the West Indies or perhaps South America. No attempt has been made to give range information beyond the Southeast, except for occasional reference to specific ranges beyond this area for a particular species that might have some special interest.

10. Photographic Credits

Photo credit for each picture is indicated by the set of three initials beneath each color picture. In addition to the main set of photographs by the coauthors, primarily B.J.T. with a few by C.R.B., it is a pleasure to give special thanks to Michael Cousens, Eve Hannahs, the late Dr. W. S. Justice, Dodie Pedlow, and Robert C. Speas, each of whom provided ten or more slides to the final collection, and to acknowledge the excellent photographs by the following people:

Oren L. Bass	O. L. B.	Michael I. Cousens	M. I. C.
Bradley C. Bennett	B. C. B.	Cecil C. Frost	C. C. F.

Rob K. Gardner	R. K. G.	Loyal A. Meyerhoff	L. A. M.
Kristen P. Giebel	K. P. G.	Anna M. Mueller	A. M. M.
Eve A. Hannahs	E. A. H.	Dodie N. Pedlow	D. N. P.
Robin B. Huck	R. B. H.	Robert K. Peet	R. K. P.
William S. Justice	W. S. J.	George C. Pyne	G. C. P.
Steven W. Leonard	S. W. L.	Robert C. Speas	R. C. S.
Anne H. Lindsey	A. H. L.	Henry O. Whittier	H. O. W.

11. Key Character Summary Code

Botanically, plants are classified on the basis of a series of more or less technical characteristics. However, many of the important plant characteristics that are used in botanical classification are relatively nontechnical (e.g., leaf arrangement, petal number, or plant habit), are easily observed, and thus are very useful in plant identification. In order to assist those who may wish to narrow their choices when trying to find the name of a particular plant (rather than look through all 500 pictures each time!), and also learn a bit more about plant family characteristics, a brief but informative nine character summary is given for each species. This summary usually contains some botanical information not mentioned in the text and can be very helpful in identification to the family level, or beyond, for most of the 500 plants included in this book. The summaries are, of necessity, somewhat generalized, but with careful observation and a little practice with "interpretation," the character summary code will make your field trips more rewarding and your identifications more accurate.

The various botanical characters involved in the code are illustrated, with their code letter or number, inside of the front and back covers. Before trying to "key out" an unknown plant (see keys at end of text) it would be helpful to compare the features of the plant you have with the series of illustrated key characters to be sure you have the basic information that will be needed for proper identification. Note that the first three code elements are important general plant characters, the second set of code elements refers to three important leaf characteristics and the final three elements of the code provide important botanical information about the flower. In special cases certain of the entries have been somewhat generalized to help increase the accuracy of specimen identification without undue reliance on botanical technicalities. Reference to the comparison, given below, of the codes for Sweet Pitcher Plant (137) and Cattail (78) will help illustrate both code structure and code content.

Sweet Pitcher Plant H-5B/SLE/SRR			*Cattail* A-OB/SGE/IAB		
I. *Important General Key Characters*					
H-5B/ =	plant herbaceous	(H)	A-OB/ =	plant aquatic	(A)
	petals 5	(5)		petals absent or number not known	(O)
	leaves basal	(B)		leaves basal (or appearing so	(B)
II. *Important Key Leaf Characters*					
SLE/ =	leaves simple	(S)	SGE/ =	leaves simple	(S)
	leaves linear or elongate	(L)		leaves grass-like	(G)
	leaf margins entire	(E)		leaf margins entire	(E)
III. *Important Key Flower Characters*					
SRR =	flower solitary	(S)	IAB =	flowers in a spike	(I)
	flower regular, radial	(R)		flower shape unknown	(A)
	flower color red	(R)		flower color brown	(B)

12. For Further Reference

The majority of the plants illustrated are included either in the *Flora of Tropical Florida* or in the *Manual of the Vascular Flora of the Carolinas*, or sometimes both. To assist those who may wish more specific botanical information about a certain plant we have included the family, genus, and species number (*not* page number) for each species as treated in one or both of these manuals. The abbreviation "L" is used for the Long and Lakela volume on tropical Florida and "R" is used to indicate the Carolinas work by Radford, Ahles and Bell. For additional information on *Sabal palmetto*, for example, the references are: (L 34-5-3; R 31-2-2). With this, one can quickly look up family 34 (Arecaceae), genus 5 (*Sabal*) and species 3 (*S. palmetto*) in Long and Lakela or family 31, genus 2, species 2 in the *Flora of the Carolinas* for a more detailed treatment of this species as it occurs in south Florida and in the Carolinas.

16. Season of Bloom

In northern Florida, both native and introduced plants maintain a temperate seasonality, although a given season, and especially spring, may start rather early in the year and may also be a bit prolonged. For plants in this area, the season of bloom symbol (Figure 2) will often be more helpful than for those plants growing in the essentially frost-free areas

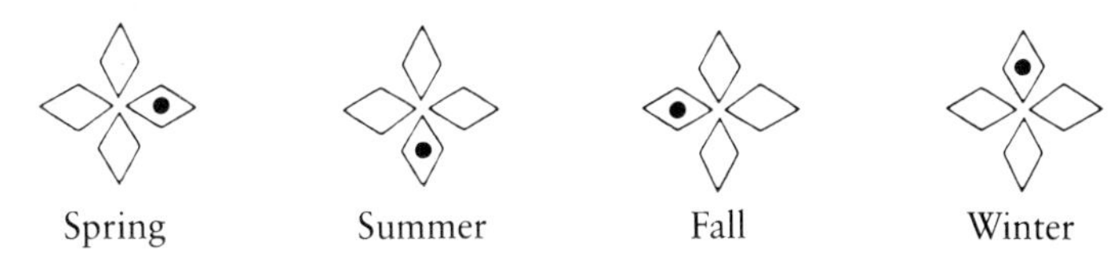

FIGURE 2. Season of bloom indicator. From left to right the dots indicate spring, summer, fall, winter.

of peninsular Florida. Even in the more tropical areas, however, most plants have a relatively specific primary season of bloom that can be indicated with confidence. Such bloom dates, though specific to season rather than to month, do provide one more piece of information of value in the identification of a particular species, but in Florida's mild climate there may be frequent exceptions!

SELECTED REFERENCES

Baker, M. F. 1926. *Florida Wild Flowers.* 245 pp. MacMillan, New York.

Cronquist, A. 1980. *Vascular Flora of the Southeastern United States.* Vol. 1. Asteraceae. University of North Carolina Press, Chapel Hill.

Godfrey, R. K. and J. W. Wooten. 1979, 1981. *Aquatic and Wetland Plants of Southeastern United States* (2 vol.). University of Georgia Press, Athens.

Kartesz, J. T. and Rosemarie Kartesz. 1980. *A Synonymized Checklist of the Vascular Flora of the United States, Canada and Greenland*, Vol. 2 The Biota of North America. 498 pp. + xlviii. University of North Carolina Press, Chapel Hill.

Long, R. W. and O. Lakela. 1971. *A Flora of Tropical Florida.* 962 pp. University of Miami Press, Coral Gables.

Luer, C. A. 1972. *The Native Orchids of Florida.* New York Botanical Garden. 293 pp.

Morton, J. F. 1974. *500 Plants of South Florida.* 163 pp. E. A. Seeman Publ., Miami.

Radford, A. E., H. E. Ahles, C. R. Bell. 1968. *Manual of the Vascular Flora of the Carolinas.* 1183 + lxi pp. University of North Carolina Press, Chapel Hill.

Small, J. K. 1933. *Manual of the Southeastern Flora.* 1554 pp. University of North Carolina Press, Chapel Hill.

Ward, D. B. (Ed.) *n.d. Rare and Endangered Biota of Florida.* Vol. 5. Plants. 175 pp. University Presses of Florida, Gainesville.

Wunderlin, Richard P. 1981. *Guide to the Flora of Central Florida.* (unpublished manuscript).

Florida Wild Flowers

and Roadside Plants

CRB

3. Mosquito Fern; Azolla

Azolla caroliniana Willdenow

The individual plants of this interesting, free-floating, aquatic fern are about 1 cm wide. The bilobed leaves may be either green or dark red.

Azolla is widespread in lakes, ponds, and slow streams throughout Florida and over much of the eastern United States and tropical America; it often forms a solid but easily broken mat on the water surface.

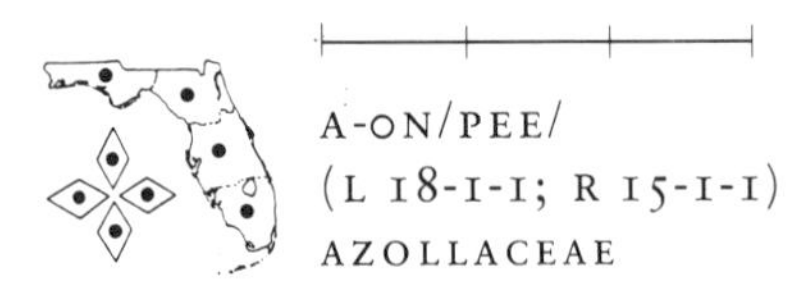

A-ON/PEE/
(L 18-1-1; R 15-1-1)
AZOLLACEAE

MIC

4. Netted Chain Fern

Woodwardia areolata (L.) Moore

The wide, green, pinnately dissected sterile, or non-spore-producing, fronds of this *Woodwardia* may be 6 dm tall. The very slender linear segments of the distinct fertile, or spore-bearing, fronds may be a bit taller.

This fern is common throughout Florida and the Southeast in bogs and wet pinelands and along swamp margins and wet ditches.

H-OB/PNE/
(L 13-2-0; R 12-1-2)
BLECHNACEAE

5. Boston Fern

Nephrolepis exaltata (L.) Schott

The clustered fronds of this handsome fern are elliptic in outline and may be a meter tall. The round sori, or clusters of sporangia, are between the midrib and margin on the underside of the leaflets.

This tropical fern is often cultivated and is widely established, often on the trunks of Cabbage Palms in their range, in hammocks, swamp margins, and even on roadsides in all sections of Florida.

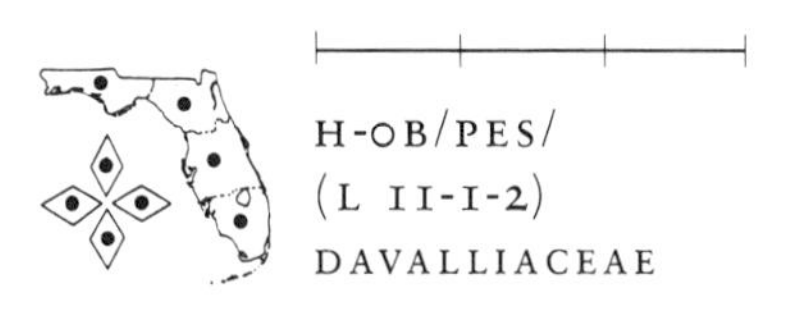

H-OB/PES/
(L 11-1-2)
DAVALLIACEAE

BJT

6. Bracken Fern

Pteridium aquilinum (L.) Kuhn

The stiff, erect, bipinnate or tripinnate leaves, triangular or ovate in outline, may be a meter or more tall and 1 meter broad. The sori are marginal on the underside of the small, linear leaflets. The leaves may be widely spaced along the perennial rhizome, or the plants may form dense thickets.

This weedy cosmopolitan fern is common in old fields, dry woodlands, sand scrub, and sandy roadsides throughout Florida.

H-OB/BOE/
(L 12-4-1; R 10-5-1)
DENNSTAEDTIACEAE

RKG

7. Cinnamon Fern

Osmunda cinnomomea Linnaeus

The cluster of succulent early spring "fiddleheads" of the Cinnamon Fern produces two kinds of leaves or fronds: a few slender, erect, cinnamon-colored spore-bearing leaves that soon wither and disappear and the broad, pinnate, dark green foliage leaves, a meter or more tall, that make this fern so attractive.

This *Osmunda* is frequent in low woods, swamps, and along pond or stream margins from south Florida to Texas and New England.

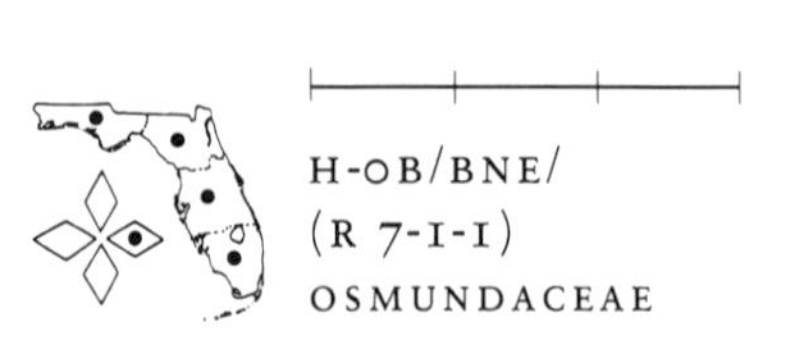

MIC

8. Royal Fern

Osmunda regalis Linnaeus

The upper leaflets of the large bipinnate fronds of the Royal Fern are greatly reduced and are fertile—that is, they bear the sporangia, or spore cases, which produce the spores of these plants. The green, linear, sterile leaflets are 4–5 cm long.

This rhizomatous perennial, also found in swamps and other wet habitats, occurs in all parts of Florida and much of the Southeast but is somewhat less frequent than the Cinnamon Fern.

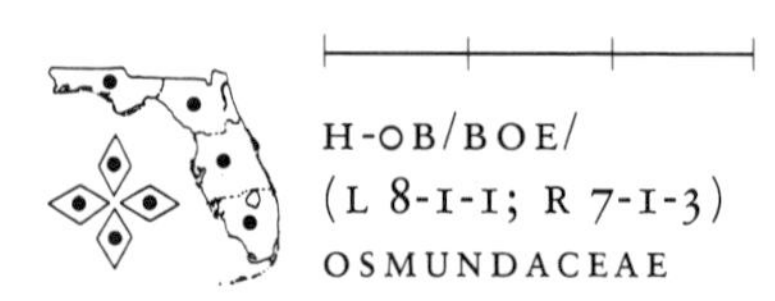

BJT

9. Resurrection Fern

Polypodium polypodioides (L.) Watt

The leathery evergreen leaves of this rhizomatous epiphyte are 1–3 dm long. They curl when dry, making the plant appear dead, but a little rain causes the leaves to reopen and thus to appear "resurrected."

Widespread throughout the Southeast, this polypody is a frequent inhabitant of the larger branches of Cypress and Live Oak or other hardwood trees where, in Florida, it often grows with various species of *Tillandsia*.

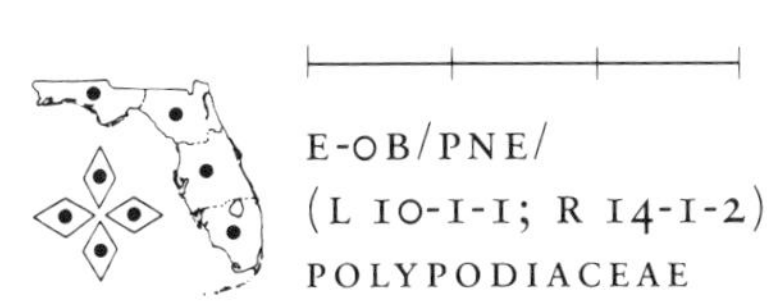

E-OB/PNE/
(L 10-1-1; R 14-1-2)
POLYPODIACEAE

MIC

10. Climbing Fern

Lygodium japonicum (Thunb.) Swartz

The vinelike fronds, or leaves, of the Climbing Fern grow all year and may reach a length of 2 meters or more. The individual, pinnately arranged, triangular leaflets are 1–2 cm long and have a toothed margin. (Our native, and less frequent, *L. palmatum* has palmate leaflets with rounded margins.)

This weedy Asiatic fern is well established along roadsides and woodland margins at many localities throughout Florida; it also occurs, but less frequently, in adjacent states.

V-OO/POD/
(L 7-2-1; R 8-1-2)
SCHIZAEACEAE

RBH

11. Florida Arrowroot; Coontie *

Zamia pumila Linnaeus

The short, woody stem and rootstock of this primitive fernlike plant (once used as a starch source by the Indians) is almost completely underground and produces a terminal crown of stiff, evergreen, pinnate leaves up to 1 meter long. The brown, fleshy, erect, female or seed-bearing cones, 10–15 cm long, are pendent when mature.

Coontie is sporadic in pinelands and hammocks throughout nearly all peninsular Florida and the Keys; it also occurs in the West Indies.

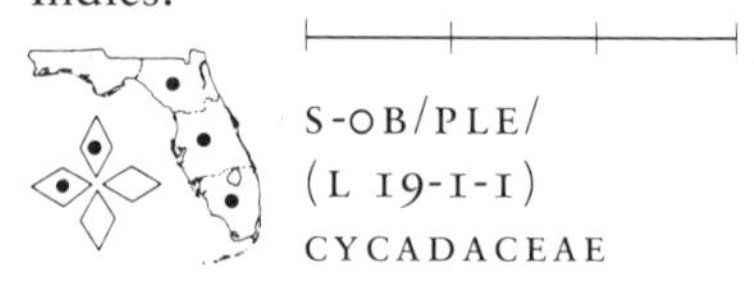

CRB

12. White Cedar

Chamaecyparis thyoides (L.) Britton, Sterns & Poggenberg

White Cedar, a small to large evergreen tree with scalelike leaves, is monoecious, that is, it produces male or pollen cones and female or seed cones on the same tree. The round brown seed cones are leathery or woody, and 5–7 mm in diameter.

Rare in acid swamps and along streams in central and north Florida, more frequent in the western Panhandle and up the Atlantic coast to the Carolinas.

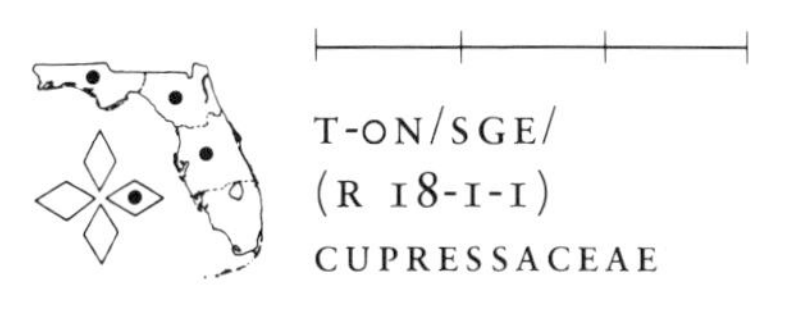

T-ON/SGE/
(R 18-1-1)
CUPRESSACEAE

CRB

13. Red Cedar

Juniperus virginiana Linnaeus

This evergreen tree, with small scalelike leaves and aromatic red heartwood often used for fence posts and for cedar chests, is dioecious: it bears male or pollen cones on one tree and round, fleshy, blue, female or seed cones on another. (The similar *J. silicicola*, Southern Red Cedar, occurs in our area south to below Tampa.)

Red cedar is common in old fields and dry open woods of much of the eastern United States but occurs in Florida only along the Georgia border.

T-ON/SGE/
(L 22-1-0; R 18-3-2)
CUPRESSACEAE

CRB

14. Long Leaf Pine

Pinus palustris Miller

Although the dark green needle-like leaves, which may be 3–4 dm long, and the conspicuous female or seed cones of this large tree are known to everyone, the small cluster of male or pollen cones at the tip of each branch are seldom seen. (The needles of Slash Pine—*P. elliottii*, the dominant timber pine throughout Florida—may be 2.5 dm long.)

This large and commercially valuable tree of the southeastern coastal plain occurs in old fields, pine flatwoods, and on sand ridges as far south as the Lake Okeechobee area.

T-OW/SGE/
(L 20-1-2; R 16-1-2)
PINACEAE

MIC

15. Stinking Cedar; Gopherwood *

Torreya taxifolia Arnott

The flat, rigid, dark green leaves and the 2–3 cm round fleshy "cones" on female trees are characteristic of this handsome evergreen now generally threatened by a fungal disease.

This rare endemic, known only from one county in Georgia and three counties along the Apalachicola River in west Florida, may be seen at Torreya State Park, where another rare gymnosperm, *Taxus floridana* or Florida Yew, also occurs.

T-OA/SLE/
TAXACEAE

WSJ

16. Bald Cypress

Taxodium distichum (L.) Richards

This large semiaquatic tree with fernlike foliage, a buttressed trunk, and woody "knees" when growing in water or saturated soil can be 30–40 meters tall and 2–3 meters in diameter. Because of the exceptional value of the timber, such majestic trees are quite rare.

Small trees are common in swamps or lakes and along streams throughout Florida and are also found along the coasts or major coastal plain rivers to the Carolinas, Illinois, and Texas.

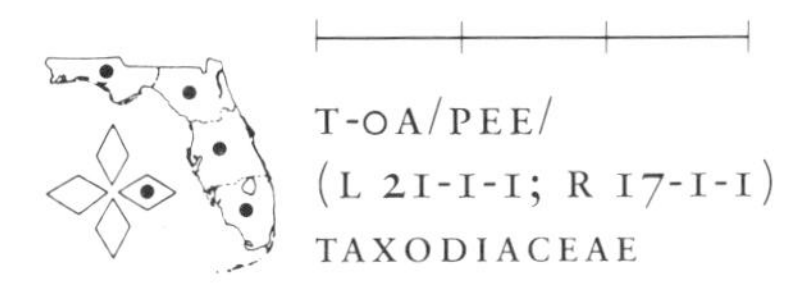

T-OA/PEE/
(L 21-1-1; R 17-1-1)
TAXODIACEAE

BJT

17. Duck Potato

Sagittaria falcata Pursh

These native aquatic perennials with elliptic leaf blades have flowers 2–3 cm across. (The leaves of the very similar *S. graminea* are usually linear or grasslike; the related and more common *S. lancifolia* may reach a height of 2 meters in some areas.)

The various species are frequent in marshes, swamps, ponds, and wet ditches from south Florida north through much of the eastern United States.

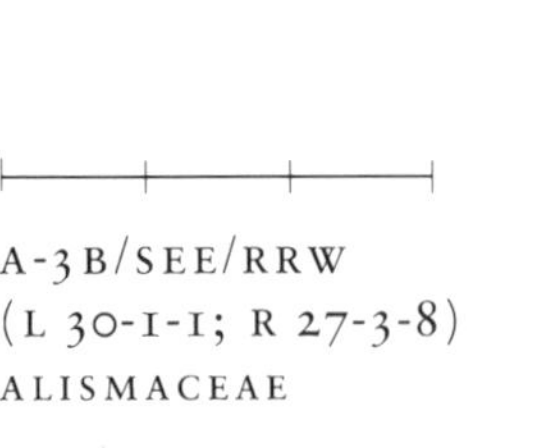

A-3B/SEE/RRW
(L 30-1-1; R 27-3-8)
ALISMACEAE

RCS

18. Arrowhead

Sagittaria latifolia Willdenow

The leaves of this rhizomatous perennial are sagittate, or arrowhead shaped, and 10–20 cm wide, but the leaves of some other species of *Sagittaria* are much narrower and lack the large, characteristic basal lobes. All species, however, have underground stems that contain stored starch and are often eaten by ducks.

These native aquatics grow in the shallow water of ponds and wet ditches from south Florida throughout much of the United States.

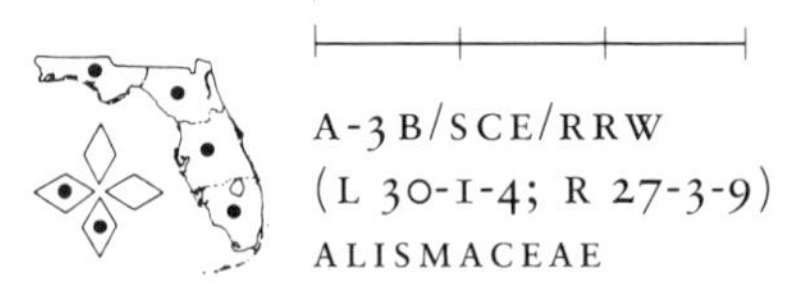

A-3B/SCE/RRW
(L 30-1-4; R 27-3-9)
ALISMACEAE

WSJ

19. Brazilian Elodea

Egeria densa Planchon

The flowers of this submersed aquatic perennial are borne just above the surface of the water on a short stalk, or pedicel. The whorled leaves, 2–3 cm long, are smooth to the touch. (Florida Elodea, *Hydrilla verticillata*, has smaller, rough leaves.)

Egeria, a popular aquarium plant from Brazil, is naturalized in clear springs and streams of central and northern Florida, and much of the southeastern United States, where heavy infestations may become a problem.

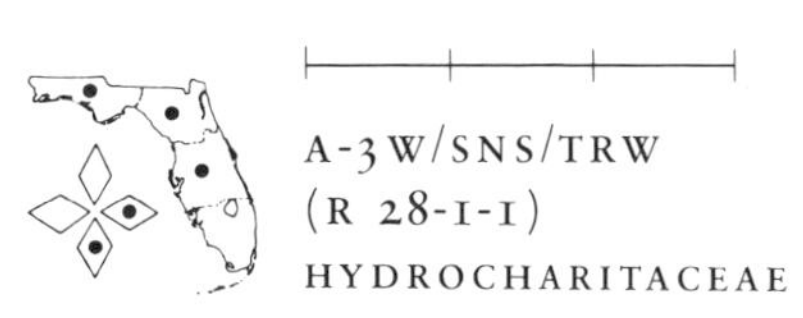

A-3W/SNS/TRW
(R 28-1-1)
HYDROCHARITACEAE

RCS

20. Colic Root

Aletris farinosa Linnaeus

Although the farinose, or mealy, flowers are only 6–8 mm long, the slender spikes are often 6 dm tall and are quite showy in large populations. (The corolla lobes of the very similar *A. obovata*, of northern Florida, are not spreading.)

These perennial herbs are frequent in savannas, prairies, and moist open areas throughout our state and the Southeast.

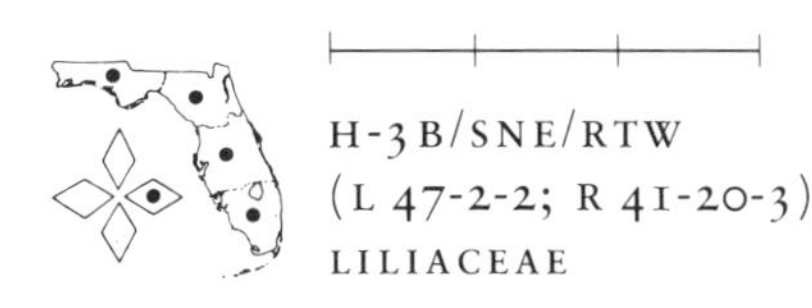

H-3B/SNE/RTW
(L 47-2-2; R 41-20-3)
LILIACEAE

BJT

21. Yellow Colic Root

Aletris lutea Small

The flowers of *A. lutea* are 6–10 mm long, a bit longer than those of *A. farinosa*. Both species have a characteristic basal rosette of narrow yellowish green leaves that are often conspicuous in the winter.

Yellow Colic Root, also a perennial, is especially frequent in the pine flatwoods of the Panhandle but also occurs in the other areas of Florida and into Louisiana and Georgia.

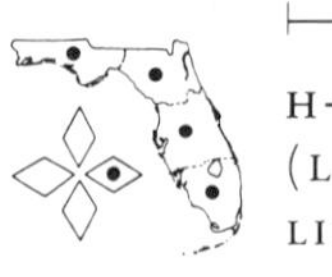

H-3B/SNE/RTY
(L 47-2-1)
LILIACEAE

BJT

22. Wild Onion

Allium canadense Linnaeus

The flowers of this pungent perennial are 1–2 cm broad and are often intermixed with small bulbils that enable these plants to propagate asexually, or vegetatively.

Often a troublesome weed of lawns and gardens over much of the eastern United States, Wild Onion reaches its southern limit near Ocala, where it is rare.

H-3B/SGE/URW
(R 41-35-2)
LILIACEAE

CRB

23. Devil's Bit; Blazing Star

Chamaelirium luteum (L.) Gray

The plants of Devil's Bit are dioecious—one plant bearing only female, or pistillate, flowers that will produce fruit and another bearing only male, or staminate, flowers that provide the pollen. The 3–5 dm tall spikes of staminate flowers are the most showy.

These perennials grow in rich deciduous forests over much of the eastern United States but are found in Florida in only a few counties around Tallahassee.

H-3B/SLE/IRW
(R 41-17-1)
LILIACEAE

BJT

24. String Lily; Swamp Lily

Crinum americanum Linnaeus

The showy clustered flowers of this tropical perennial are very fragrant; the narrow white sepals and petals are 5–10 cm long.

As one of the common names implies, these bulbous herbs frequently grow along swamp or pond margins and creek banks of peninsular and northern Florida and along the Gulf coast to Texas.

H-3B/SGE/URW
(L 47-5-1)
LILIACEAE

GCP

25. Dogtooth Violet; Trout Lily

Erythronium umbilicatum Parks & Hardin

The nodding flowers with strongly reflexed petals 2–4 cm long and the wide, elliptic, smooth, mottled leaves immediately identify this small woodland perennial.

Trout Lily is common in northern hardwood forests but is very rare in Florida, where it apparently occurs only in a single north-central county.

H-3B/SEE/SRY
LILIACEAE

RCS

26. Spider Lily

Hymenocallis crassifolia Herbert

The thin white crown in the center of each of the clustered flowers is 5–8 cm across and makes these plants even more showy than the String Lily.

A bulbous perennial that is native to swamp margins and stream banks of northern Florida and on into Louisiana and the Carolinas.

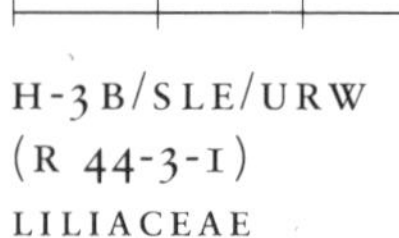

H-3B/SLE/URW
(R 44-3-1)
LILIACEAE

BJT

27. Alligator Lily

Hymenocallis palmeri Watson

This native perennial is closely related to the preceding species but differs from it by having only a single flower on each stalk, narrower leaves that are less than 1 cm wide, and a more southern distribution.

The endemic Alligator Lily is locally frequent in the prairies, glades, and cypress swamps in the area generally south of Orlando.

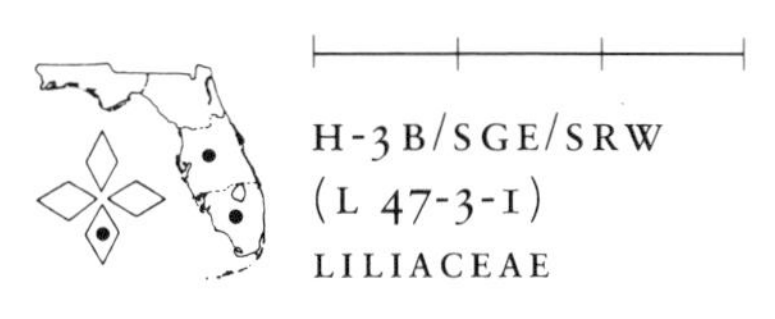

H-3B/SGE/SRW
(L 47-3-1)
LILIACEAE

BJT

28. Yellow Star Grass

Hypoxis hirsuta (L.) Coville

A low, grasslike native perennial with pubescent leaves 5–25 cm long. The short flower stalk bears three or more flowers, each 2–3 cm across. (Two very similar and closely related species—*H. juncea* and *H. wrightii*, which may not be biologically distinct—are endemic to the pine flatwoods of southern peninsular Florida.)

Widespread through the eastern United States and frequent in open woodlands and pastures of northern Florida.

H-3B/SGE/URY
(L 47-1-0; R 44-7-1)
LILIACEAE

WSJ

29. Pine Lily

Lilium catesbaei Walter

The leafy stems of this true lily may be up to 5 dm tall. The six similar and brightly colored perianth segments—three sepals and three petals—are 8–10 cm long.

The Pine Lily is frequent in the pine flatwoods and savannas throughout nearly all Florida; it is less frequent north along the coastal plain to the Carolinas.

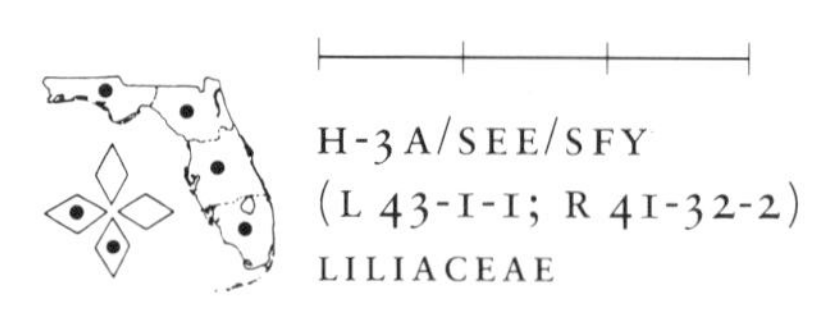

H-3A/SEE/SFY
(L 43-1-1; R 41-32-2)
LILIACEAE

WSJ

30. False Garlic

Nothoscordum bivalve (L.) Britton

Although this bulbous perennial looks much like a wild onion—and is sometimes classified as an *Allium*—it has no onionlike smell, thus the common name. The flowers are 1–2 cm in diameter. (The related African species, *N. inodorum*, with leaves more than 5 mm wide, is an occasional escape from cultivation in our area.)

False Garlic is usually a weed in lawns, pastures, and along roadsides from Lee County northward.

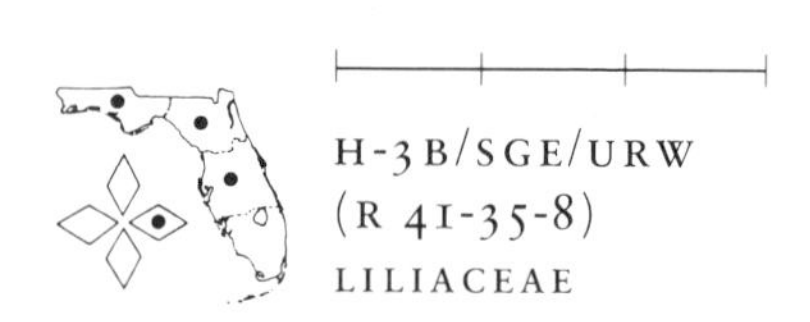

H-3B/SGE/URW
(R 41-35-8)
LILIACEAE

DNP

31. Wake Robin; Trillium

Trillium maculatum Rafinesque

The three wide mottled leaves, 1–2 dm long, and the solitary, erect, flower on each stem aid in the quick recognition and identification of this woodland perennial.

Trillium is another rare northern element of the varied Florida flora which occurs on calcareous bluffs in only a few north-central counties of the state.

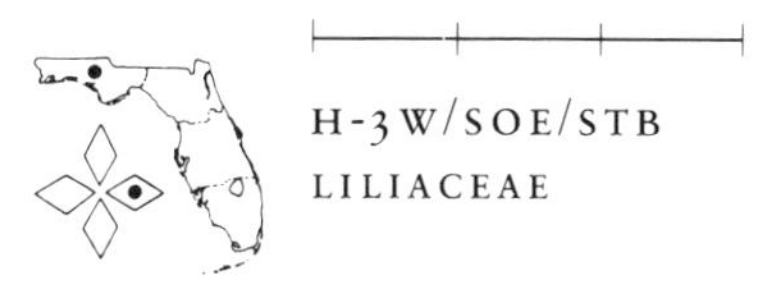

H-3W/SOE/STB
LILIACEAE

BJT

32. Bellwort

Uvularia perfoliata Linnaeus

The nodding solitary flower of this rhizomatous perennial is 2–3 cm long, and the elliptic, perfoliate leaves are 4–8 cm long. (The leaves of two other rare species, *U. floridana* and *U. sessilifolia*, also in this area, are not perfoliate.)

Bellwort is a spring ephemeral of more northern hardwood forests that reaches its southern limit in the rich woods of Gadsden and Leon counties, where it is rare.

H-3A/SEE/SFY
(R 41-33-1)
LILIACEAE

RCS

33. Atamasco Lily

Zephyranthes atamasco (L.) Herbert

The erect flowers of these low bulbous perennials are about 5 cm long and rapidly change from pure white to pink as they age. (In *Z. simpsonii* of central and south Florida the flowers are usually pink even when just open.)

A frequent native of wet meadows, glades, and low woods from central Florida to Texas and Virginia.

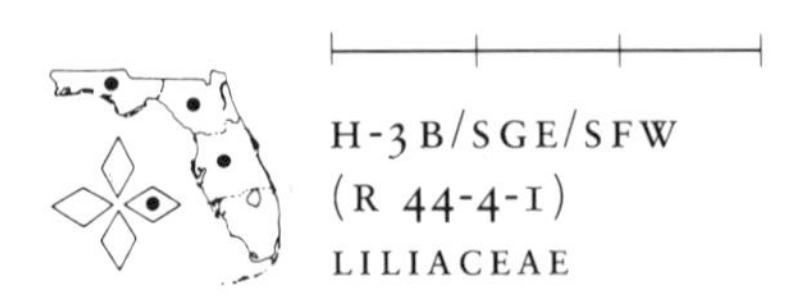

H-3B/SGE/SFW
(R 44-4-1)
LILIACEAE

RCS

34. Crow Poison

Zigadenus densus (Desr.) Fernald

The compact raceme of Crow Poison may be a decimeter or more long on a slender stalk somewhat longer than the linear basal leaves. (These plants are very similar to, and often confused with, the related Fly Poison, *Z. muscaetoxicus*, found in northern Florida.)

Frequent in pine flatwoods and savannas from central Florida north to the Carolinas and west to Mississippi.

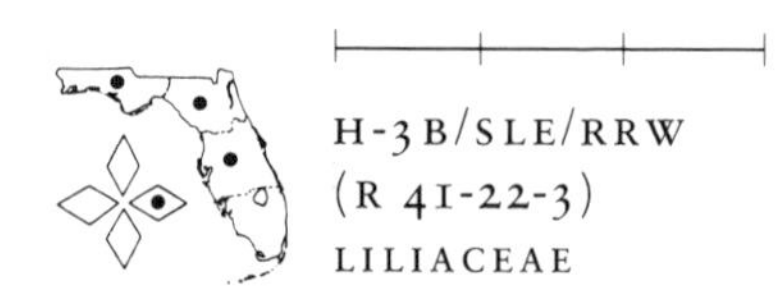

H-3B/SLE/RRW
(R 41-22-3)
LILIACEAE

WSJ

35. Snakeroot

Zigadenus glaberrimus Michaux

The six perianth segments, three sepals and three petals, are 1–1.5 cm long and have two distinct glands at the base. The larger flowers, and the more open raceme, easily identify this species as compared with *Z. densus.*

A relatively frequent perennial of northern Florida flatwoods and much of the southeastern coastal plain, Snakeroot often forms large clumps by vegetative reproduction.

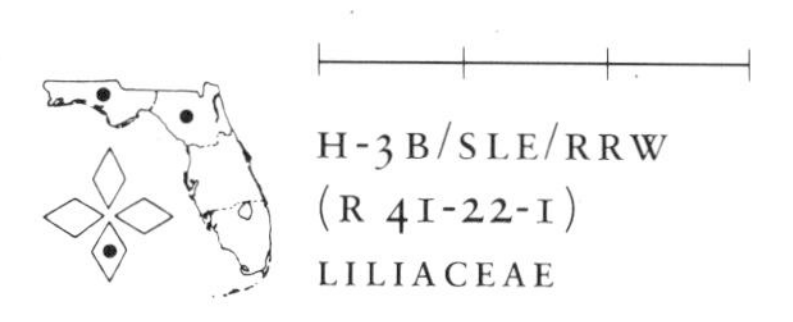

H-3B/SLE/RRW
(R 41-22-1)
LILIACEAE

WSJ

36. Manfreda

Manfreda virginica (L.) Rose

This relative of the western desert Century Plant has a basal rosette of fleshy lanceolate leaves, 8–40 cm long, which are often spotted with purple. The inconspicuous racemose flowers, on stalks up to 1.5 meters tall, are very fragrant.

An infrequent perennial herb of dry or rocky deciduous woods from west Florida to Texas and North Carolina.

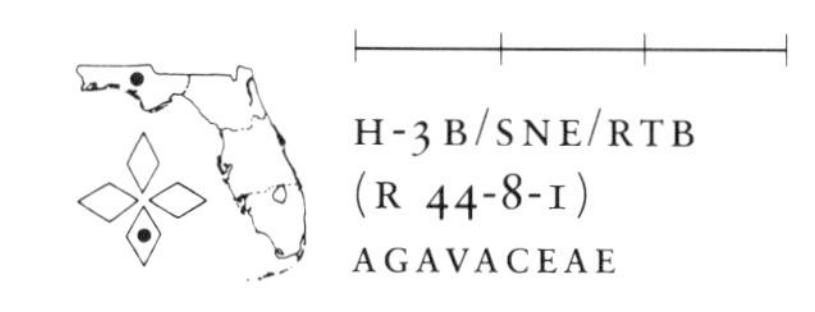

H-3B/SNE/RTB
(R 44-8-1)
AGAVACEAE

BJT

37. Spanish Bayonet

Yucca aloifolia Linnaeus

The erect, treelike trunks of these rhizomatous perennials may be up to 1.5 meters tall with the upper portion covered by long, stiff, sharply tipped evergreen leaves. The individual flowers are 4–6 cm long.

Frequent as an escape from cultivation or persisting in sandy woods and waste areas, this native of Mexico is established along the Florida coasts, and less frequently, inland.

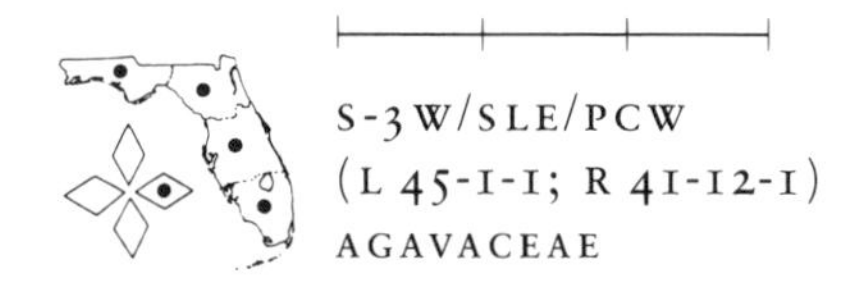

S-3W/SLE/PCW
(L 45-1-1; R 41-12-1)
AGAVACEAE

CRB

WSJ

38. Yucca

Yucca filamentosa Linnaeus

The leaves of this species of *Yucca* are all basal, are not stiff, and the margins fray into coarse fibers or filaments. The individual flowers are 3–6 cm long.

This native perennial is found on sandhills and in dry, open deciduous woods from south Florida north and west over much of the southeastern United States.

H-3B/SLE/PCW
(R 41-12-3)
AGAVACEAE

39. Redroot

Lachnanthes caroliniana (Lam.) Dandy

The flat, linear, basal leaves of this perennial are usually 0.5–1.5 cm wide and 2–5 dm long. The compact, branched, tomentose inflorescence expands somewhat as the flowers mature. The red sap of the roots and rhizomes accounts for the common name.

Common to frequent in pine flatwoods, savannas, and wet ditches throughout Florida and much of the eastern coastal plain.

H-3B/SLE/URW
(L 46-1-1; R 45-1-1)
HAEMODORACEAE

WSJ

40. Goldcrest

Lophiola aurea Ker-Gawler

The small flowers of Goldcrest have six stamens as opposed to the three found in Redroot; also, the inflorescence of this species is more open and the leaves are narrower, generally 5 mm or less in width.

Infrequent, but sometimes locally abundant in marshes, bogs, wet ditches, and flatwoods of northern Florida and the coastal plain of the Southeast.

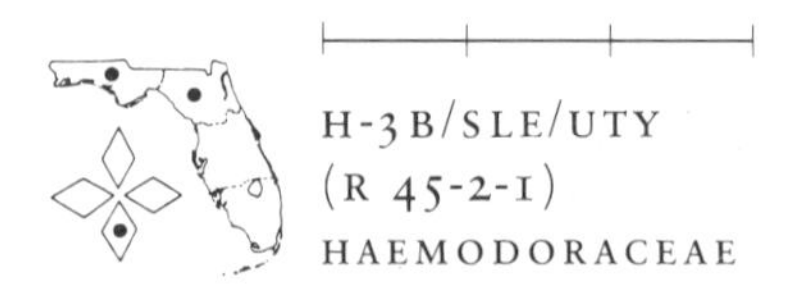

H-3B/SLE/UTY
(R 45-2-1)
HAEMODORACEAE

CRB

41. Bamboo Vine; Catbrier

Smilax laurifolia Linnaeus

The alternate, evergreen leaves of this usually thorny, woody vine are 5–15 cm long. The axillary clusters of dark blue or black fruits, which take 18 months to ripen, are more conspicuous than the small umbels of greenish flowers. (*S. walteri*, of northern Florida, has red fruits.)

A common climbing or trailing vine of savannas and dry pinelands as well as coastal dunes throughout Florida and much of the Southeast.

V-3A/SEE/URG
(L 44-1-1; R 41-2-10)
SMILACACEAE

KPG

42. Air Yam

Dioscorea bulbifera Linnaeus

In Florida this introduced, tropical climbing vine does not flower but reproduces only vegetatively by the characteristic aerial tubers that may be from 1–10 cm in diameter. The starchy tubers are edible after cooking. The cordate, or heart-shaped, leaves are 5–25 cm long and equally as wide.

Air Yam is an aggressive weed in many areas of south and central Florida where it grows in thickets, waste areas, and hedge or fence rows.

V-OA/SCE/---
(L 48-1-1)
DIOSCOREACEAE

WSJ

43. Purple Flag; Iris

Iris tridentata Pursh

The sepals of this Iris are 6–8 cm long and, as with Southern Blue Flag, are larger than the petals; the richer color of the flower may aid in its recognition.

Infrequent to locally abundant in pine flatwoods and along pond margins of northern Florida, and northward into the Carolinas.

H-3B/SLE/STB
(R 46-5-4)
IRIDACEAE

BJT

44. Southern Blue Flag

Iris virginica Linnaeus

The three sepals of the Southern Blue Flag are 7–10 cm long and are larger and more colorful than the three erect petals. (The similar *I. hexagona*, found throughout Florida, has a zig-zag stem; Dwarf Iris, *I. verna*, is known from only a few northern counties.)

This rhizomatous perennial can be found in swamps and wet prairies of northern Florida and into neighboring states to Virginia and Texas.

H-3B/SLE/STB
(R 46-5-2)
IRIDACEAE

EAH

45. Celestial Lily *

Nemastylis floridana Small

The flowers of this low, fall-blooming perennial are 2–3 cm across and are reported to be open for only a few hours in the late afternoon.

A rare endemic of a few scattered swamps and marshes of peninsular and north Florida.

H-3B/SGE/SRB
IRIDACEAE

BJT

46. Blue-eyed Grass

Sisyrinchium atlanticum Bicknell

These slender-leaved perennials often grow in clumps or tufts that are easily mistaken for a grass when the plant is not in flower. The five or six species in Florida are all generally similar in appearance and usually produce several flowers, one at a time, from a single spathe.

One or more of the various species grows in each region of Florida in a variety of habitats: flatwoods, pastures, wet fields, disturbed areas, and scrub. This species, *S. atlanticum*, is frequent in flatwoods throughout the state and much of the Southeast.

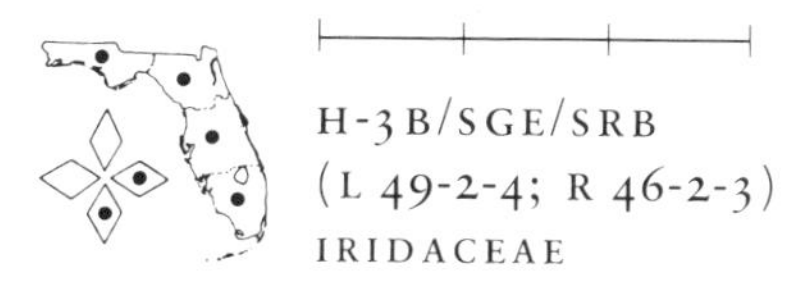

H-3B/SGE/SRB
(L 49-2-4; R 46-2-3)
IRIDACEAE

BJT

47. Blue-eyed Grass

Sisyrinchium rosulatum Bicknell

The pale blue to white flowers of this low, spreading annual are about 1 cm in diameter and have a distinctive yellow eye inside a maroon band.

These introduced plants, known from several other southeastern states, may be found sporadically in lawns, pastures, and along open roadsides of central and northern Florida.

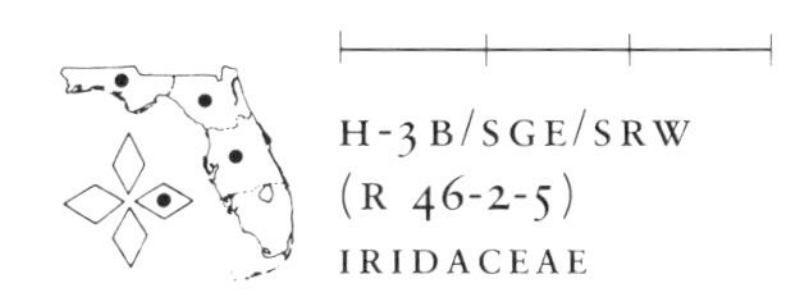

H-3B/SGE/SRW
(R 46-2-5)
IRIDACEAE

OLB

48. Spider Orchid*

Brassia caudata (L.) Lindley

The long, slender "tails" of the lateral sepals of these striking tropical orchids may be 10–15 cm long and give the flowers their common name. The stout, strap-shaped leaves, up to 5 cm wide, may be 4 dm long.

These epiphytic orchids are very rare on trees in a few hammocks at the southern tip of Florida, the northern limit of its range.

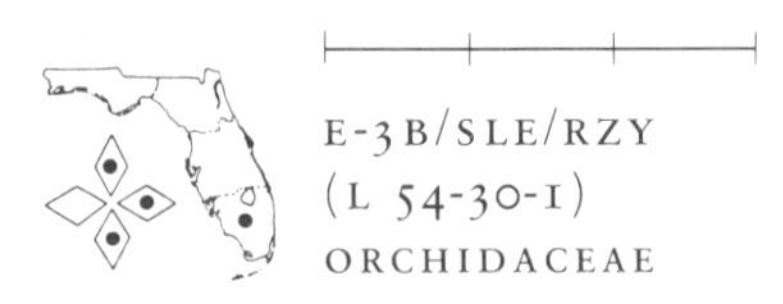

E-3B/SLE/RZY
(L 54-30-1)
ORCHIDACEAE

BJT

49. Pale Grass Pink *

Calopogon pallidus Chapman

The 3-cm-long flowers of Pale Grass Pink, with strongly reflexed lateral sepals, may be white to pale or dark pink and are often occupied by spiders that feed on the insects attracted to the blooms.

These terrestrial perennials are found in open marshy meadows and savannas throughout Florida and along the coast to Louisiana and the Carolinas.

H-3B/SLE/RZW
(L 54-6-2; R 49-10-3)
ORCHIDACEAE

BJT

50. Grass Pink *

Calopogon tuberosus (L.) Britton, Sterns & Poggenberg

This is the largest of our Grass Pinks with flowers 3–5 cm long. (Two other pink-flowered Florida species, *C. barbatus* and *C. multiflorus*, have smaller but more numerous flowers.)

This terrestrial orchid, widespread but only locally frequent over much of the eastern United States, is found throughout Florida in moist open pinelands and prairies.

H-3B/SLE/RZR
(L 54-6-1; R 49-10-2)
ORCHIDACEAE

BJT

51. Thickroot Orchid*

Campylocentrum pachyrrhizum (Reichenb.) Rolfe

This diminutive, epiphytic orchid may produce a few fleshy leaves, but most of its food is manufactured by the flat, green, leaflike roots. The small, pendulous spike of flowers is only 2–4 cm long.

Known primarily from the islands of the Caribbean and the northeast coast of South America, this tropical perennial occurs in the United States only in the Fahkahatchee Swamp of Collier County.

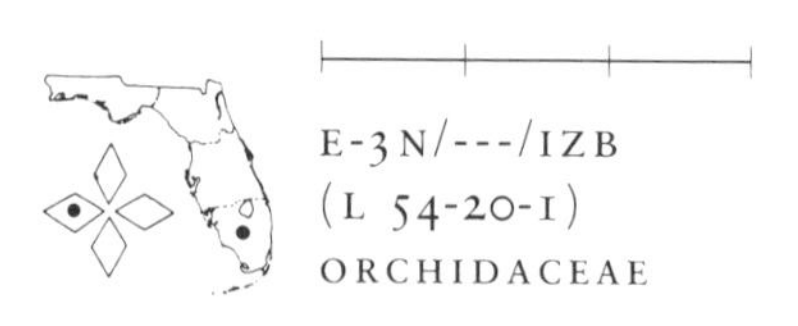

E-3N/---/IZB
(L 54-20-1)
ORCHIDACEAE

BJT

52. Rosebud Orchid*

Cleistes divaricata (L.) Ames

Each stalk of the Rosebud Orchid bears a single striking flower 6–8 cm across, and a solitary elliptic or lanceolate leaf 5–20 cm long.

This rare terrestrial perennial of low pinelands of northern Florida also occurs along the coasts to Texas and Maryland.

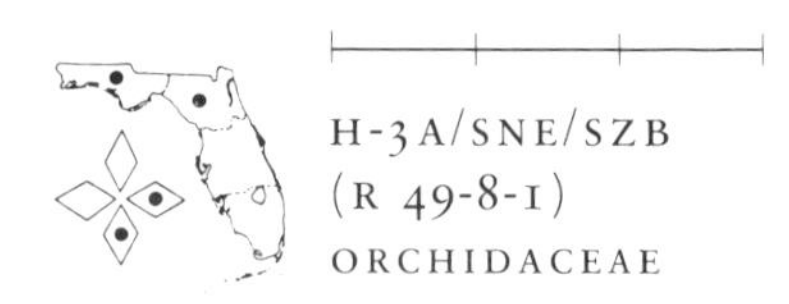

H-3A/SNE/SZB
(R 49-8-1)
ORCHIDACEAE

WSJ

53. Spring Coral Root*

Corallorhiza wisteriana Conrad

Coral Root produces a loose raceme of small flowers, less than 1 cm long, on a leafless flowering stalk up to 4 dm tall. (The similar *C. odontorhiza*, of northern Florida, blooms in the fall.)

This saprophytic terrestrial perennial, which may be locally frequent, is found in hammocks and deciduous woods of central and northern Florida and on through much of the Southeast.

H-3N/---/RZB
(R 49-20-3)
ORCHIDACEAE

BJT

54. Shell Orchid*

Encyclia cochleata (L.) Dressler

Shell Orchids produce a flower stalk up to 5 dm long from a stout pseudobulb that also produces one to three linear leaves. The racemose flowers are 4–5 cm across.

These tropical epiphytes, known primarily from Central and South America and the islands of the Caribbean, may occasionally be found on trees in the moist hammocks of south Florida.

E-3B/SNE/RZY
(L 54-27-2)
ORCHIDACEAE

BJT

55. Dingy Epidendrum *

Epidendrum anceps Jacquin

The variable brownish yellow or yellow green flowers of this *Epidendrum* are 1–2 cm across and the leafy stems may be a meter long.

This perennial South American epiphyte, often on Royal Palm, reaches its northern limit in hammocks of the Big Cypress Swamp and the Everglades.

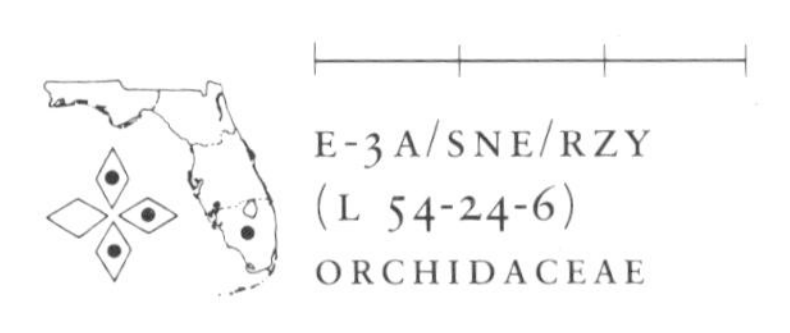

E-3A/SNE/RZY
(L 54-24-6)
ORCHIDACEAE

BJT

56. Green-fly Orchid *

Epidendrum conopseum R. Brown

The greenish white to cream flowers of the Green-fly Orchid are about 2 cm across and are produced in an open raceme.

A relatively frequent epiphyte often found with Resurrection Fern on the branches of large deciduous trees in various habitats from central Florida along the coast to Texas and the Carolinas.

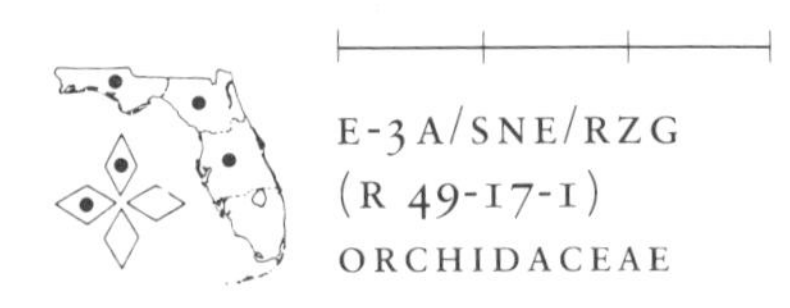

E-3A/SNE/RZG
(R 49-17-1)
ORCHIDACEAE

BJT

57. Rigid Epidendrum *

Epidendrum rigidum Jacquin

The small, rigid, greenish flowers of this epiphyte are less than 1 cm long, few in number, and are quite inconspicuous.

Primarily a species of South and Central America and the Caribbean islands, this is another *Epidendrum* that reaches its northern limit in the hammocks of south Florida.

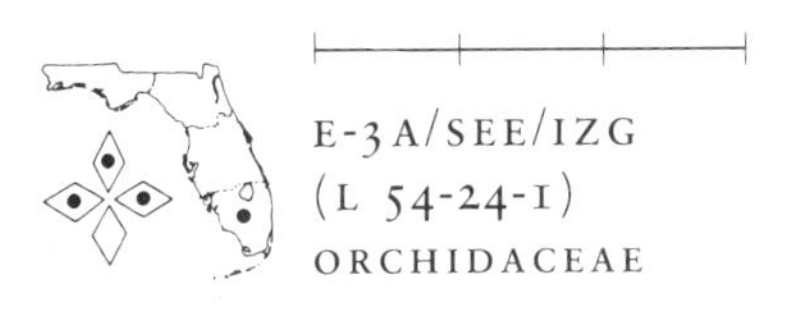

E-3A/SEE/IZG
(L 54-24-1)
ORCHIDACEAE

DNP

58. Long-horned Orchid *

Habenaria quinqueseta (Michx.) Eaton

Five slender seta, or narrow, linear, perianth lobes, and a long spur give the 2–3 cm wide flowers a rather spiderlike appearance. The erect, leafy stem is up to 5 dm tall.

An infrequent terrestrial tropical perennial of hammocks and pine flatwoods throughout Florida and along the coasts to Texas and South Carolina.

H-3A/SNE/IZW
(L 54-1-1; R 49-3-5)
ORCHIDACEAE

WSJ

59. Water-spider Orchid *

Habenaria repens Nuttall

The rather stout, erect, leafy stems of this relatively common but often overlooked orchid may be 5 dm tall and the spicate flowers about 1 cm across. (The very rare *H. distans*, of south Florida, has only basal leaves and the flowers are about 2 cm across.)

This terrestrial orchid grows in moist to wet woods, meadows, and open ditches in all areas around the Caribbean and north along the coast to the Carolinas.

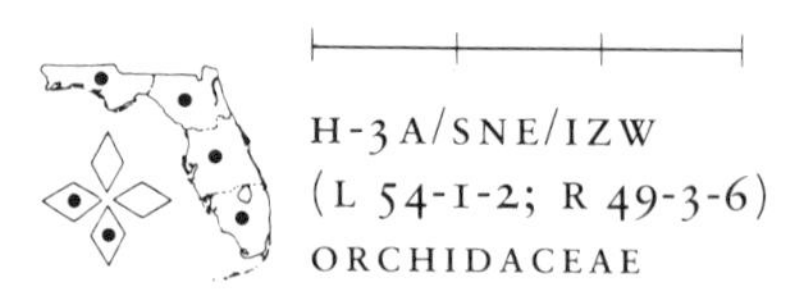

H-3A/SNE/IZW
(L 54-1-2; R 49-3-6)
ORCHIDACEAE

BJT

60. Delicate Ionopsis *

Ionopsis utricularioides (Sw.) Lindley

The delicate pinkish lavender flowers, about 1 cm across, are on a leafless scape, or flower stalk, up to 5 dm long that bears a panicle of from one to seventy blooms.

This rather rare epiphytic South American and Caribbean orchid reaches its northern limit on trees in hammocks of a few counties of south Florida.

E-3B/SEE/PZB
(L 54-31-1)
ORCHIDACEAE

MIC

61. Rose Pogonia *

Pogonia ophioglossoides
(L.) Jussieu

The one to four fragrant flowers on each slender stem of Rose Pogonia are 2.5–3 cm across and the single elliptic leaf, halfway up the stem, is about 10 cm long.

The range of this terrestrial perennial extends from Canada to Florida and Texas. It is relatively frequent in the marshes, bogs, and pine flatwoods of central and northern Florida.

H-3A/SEE/RZR
(R 49-7-1)
ORCHIDACEAE

DNP

62. White Fringed Orchid *

Platanthera blephariglottis
(Willd.) Lindley

The conspicuously fringed lip, or primary petal lobe, of each of the numerous flowers is 1–2 cm long; the two to four narrowly lanceolate leaves are reduced to bracts below the spicate inflorescence.

A terrestrial perennial of marshy, open meadows that is locally frequent from central Florida along the coasts to Texas and New Jersey.

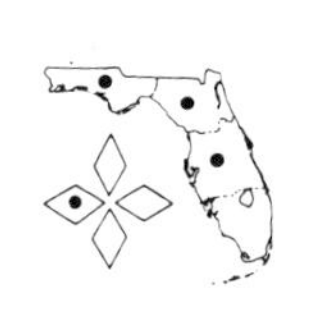

H-3A/SNE/IZW
(R 49-3-13)
ORCHIDACEAE

WSJ

63. Crested Fringed Orchid*

Platanthera cristata (Michx.) Lindley

The fringed ovate lip of this terrestrial orchid is 4–5 mm long and the numerous flowers are in a compact, almost cylindrical raceme. (The lip and flower of the very similar and more frequent *P. ciliaris*, which has the same range, are twice as large.)

Infrequent to rare in Florida meadows and pine flatwoods from near Tampa north and west to Massachusetts and Texas.

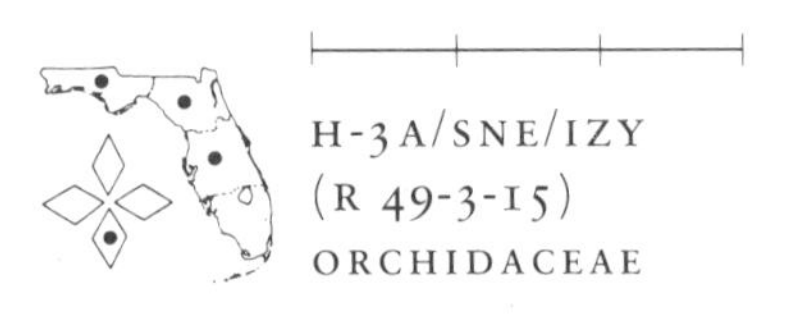

H-3A/SNE/IZY
(R 49-3-15)
ORCHIDACEAE

WSJ

64. Snowy Orchid*

Platanthera nivea (Nutt.) Luer

The smaller flowers and the absence of any fringe on the lip, which is uppermost in these flowers, quickly identify the Snowy Orchid.

This terrestrial perennial is relatively frequent in open bogs and moist meadows of all sections of Florida but is less frequent or rare northward to Maryland and west to Texas.

H-3A/SNE/IZW
(L 54-1-3; R 49-3-9)
ORCHIDACEAE

EAH

65. Scarlet Ladies' Tresses *

Spiranthes lanceolata (Aubl.) Leon

The striking red to yellow flowers of this showy plant are 2–3 cm long and the three to six elliptic or lanceolate basal leaves, which appear after flowering, are 1–4 dm long.

This tropical terrestrial orchid is rare or infrequent in south and north Florida but frequent in the pine flatwoods, open pastures, and even roadsides of central Florida.

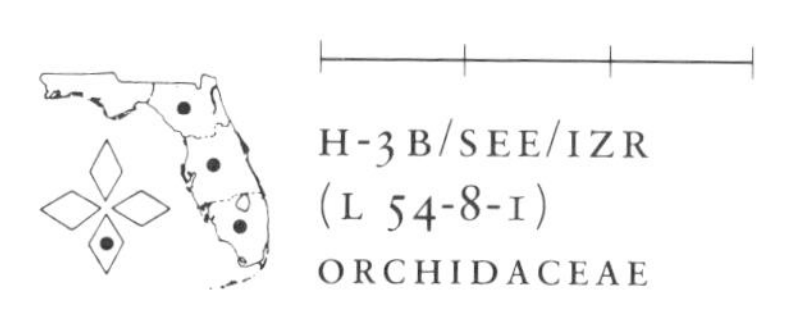

H-3B/SEE/IZR
(L 54-8-1)
ORCHIDACEAE

LAM

66. Fragrant Ladies' Tresses *

Spiranthes odorata (Nutt.) Lindley

The regularly spiraled flowers of Fragrant Ladies' Tresses are 5–15 mm long and are quite similar to those of the widespread and variable *S. cernua*, Nodding Ladies' Tresses, with which this species is also sometimes classified.

Relatively frequent in, or bordering, hardwood or cypress swamps throughout Florida and bogs, wet meadows, and stream margins along the coast to Texas and the Carolinas.

H-3B/SEE/IZW
(L 54-8-8; R 49-12-2)
ORCHIDACEAE

BJT

67. Lesser Ladies' Tresses *

Spiranthes ovalis Lindley

This delicate little orchid, with flowers only 4–6 mm long and a pubescent flower stalk 2–4 dm long, has sometimes been considered only a variety of the similar but larger, more common, and widespread Nodding Ladies' Tresses, *S. cernua*.

Rare in moist hardwood forests and along swamp margins of north and possibly west Florida, and much of the Southeast.

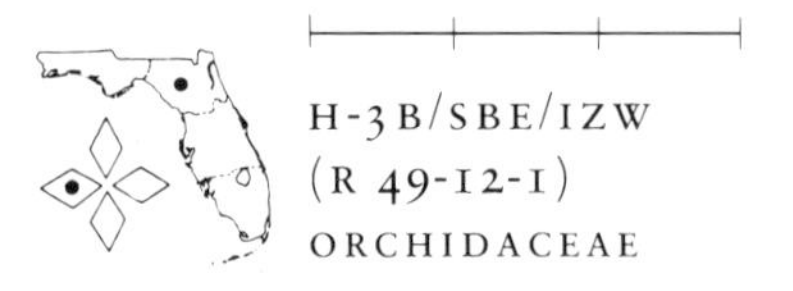

H-3B/SBE/IZW
(R 49-12-1)
ORCHIDACEAE

BJT

68. Grass-leaf Ladies' Tresses *

Spiranthes praecox (Walt.) Watson

The strong spiral twist of the 1-cm-long flowers along the spike demonstrates the origin of the generic name for these terrestrial orchids and the five or six slender leaves account for the common name. (These plants are very similar to *S. vernalis* of our area, which also blooms in early spring.)

Relatively frequent and widespread in open woods, grasslands, flatwoods, and savannas of Florida and the coastal areas of other southeastern states.

H-3A/SGE/IZW
(L 54-8-10; R 49-12-6)
ORCHIDACEAE

CRB

69. Vanilla Orchid*

Vanilla planifolia Andrews

When properly cured the slender elongate fruit pods of this tropical epiphytic vine are the source of commercial vanilla. The flowers are about 6–8 cm across and the fleshy, sessile, lanceolate leaves are up to 2 dm long.

Often cultivated and escaped or native in the hammocks and swamps of south Florida, which is its northern limit.

V-3A/SNE/RZY
(L 54-19-2)
ORCHIDACEAE

EAH

70. Green Dragon

Arisaema dracontium (L.) Schott

The large solitary leaf of this native perennial is divided into seven to fifteen segments and stands well above the "flower" with its slender greenish white spathe, 3–5 cm long.

Native but infrequent in low rich woods of central and northern Florida and much of the eastern United States.

H-OB/MNE/TAG
(R 32-5-1)
ARACEAE

CRB

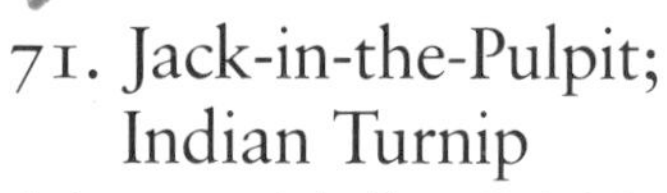

71. Jack-in-the-Pulpit; Indian Turnip

Arisaema triphyllum (L.) Schott

The colorful green to maroon or striped spathe, up to 8 cm long, forms the "pulpit" that encloses the erect, cylindrical spadix with its many minute male and female flowers. The raw corm is very pungent, but is edible when boiled. (*A. acuminatum*, with a longer spathe tip, occurs in shady hammocks of central and south Florida.)

These perennial herbs are found in rich alluvial woods chiefly in northern Florida and much of the eastern United States.

H-OB/TEE/TAG
(R 32-5-2)
ARACEAE

CRB

72. Monstera

Monstera deliciosa Liebmann

The glossy, perforated, and deeply cut ovate leaves of this dramatic introduced vine may be nearly a meter long. The leathery green cones of tasty fruits with the flavor of pineapple can be 10–20 cm long and are sometimes sold in the markets.

This popular, tropical, ornamental philodendron may sometimes persist around old homesites in south and central Florida.

V-OA/SOC/TAG
ARACEAE

BJT

73. Golden Club

Orontium aquaticum Linnaeus

The small flowers of members of the Arum family (Araceae) are borne on a specialized stem, or spadix. The yellow, terminal, fertile or flower-bearing portion of the naked spadix of Golden Club is 5–10 cm long.

These rhizomatous perennial aquatics are native to the acid bogs, stream margins, and wet ditches of northern Florida and much of the southeastern United States.

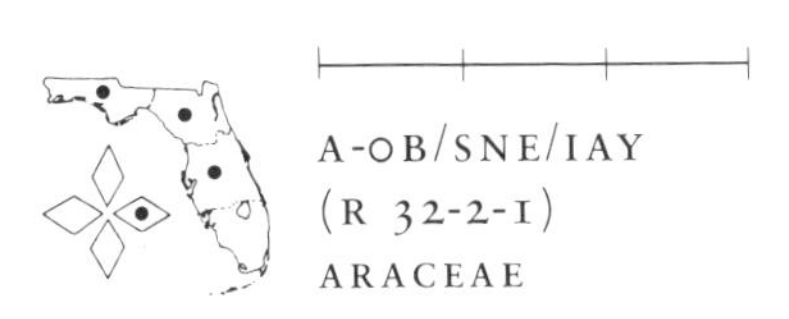

A-OB/SNE/IAY
(R 32-2-1)
ARACEAE

RCS

74. Arrow Arum; Tuckahoe

Peltandra virginica (L.) Schott

The large, glossy, sagittate, or arrowhead-shaped, leaves of this robust aquatic perennial may be from 5–20 cm wide. The green spathe surrounding the spadix and later the greenish to black fruits helps to differentiate this species from White Arum.

Tuckahoe is frequent to common in swamps, marshes, lake margins, and stream banks throughout Florida and much of the eastern United States.

A-OB/SCE/TAG
(L 35-2-1; R 32-4-1)
ARACEAE

RKP

75. Water Lettuce

Pistia stratiotes Linnaeus

The small greenish white spathes of this pantropical aquatic herb are quite inconspicuous when compared with the floating silvery green leaf rosettes, which are 5–25 cm in diameter.

Large colonies of Water Bonnets, as these plants may also be called, often completely cover quiet streams, lakes, and ponds of peninsular Florida and can block water flow and boat traffic.

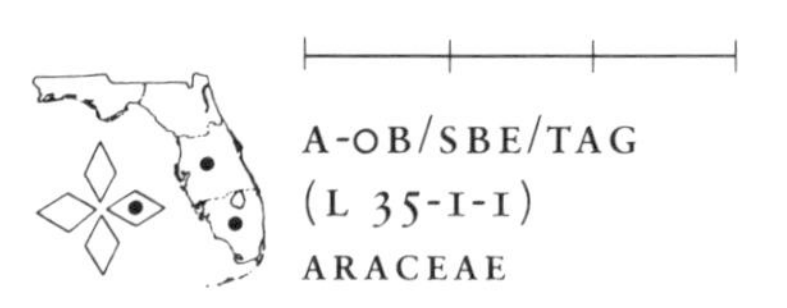

A-OB/SBE/TAG
(L 35-1-1)
ARACEAE

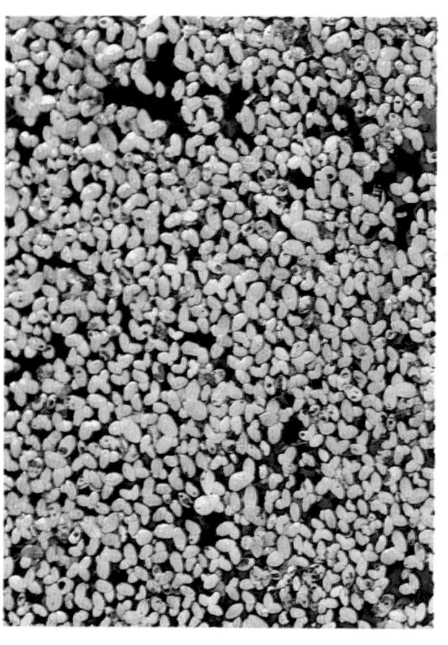

CRB

76. Duckweed

Lemna minor Linnaeus

The small obovate to elliptic plants of *Lemna* are only 2–3 mm long and are one of the smallest flowering plants in the world. The minute flowers have no sepals or petals but consist of a single stamen or a single pistil. Most reproduction is asexual.

These perennial floating aquatics (and the related *Spirodella polyrhiza*, *L. perpusilla*, and *Wolffiella floridana*) are found throughout Florida, and sporadically in other southeastern states, on rich, quiet, or stagnant freshwater pools.

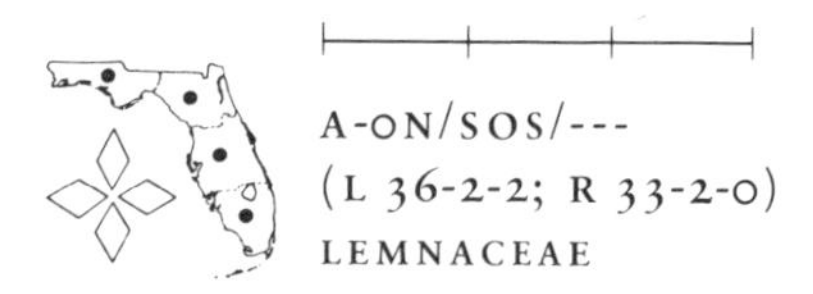

A-ON/SOS/---
(L 36-2-2; R 33-2-0)
LEMNACEAE

WSJ

77. Burr-reed

Sparganium americanum Nuttall

The larger, lower, spherical flower heads, 1.5–3 cm in diameter, bear pistillate, or female, flowers and the smaller, upper heads bear staminate, or male, flowers. The plant, often a meter tall, is an important game food.

These rhizomatous, aquatic perennials are found in lake, pond, and stream margins from the Lake Okeechobee area, where they are rare, north with increasing frequency over much of the eastern United States.

A-3A/SLE/KAG
(R 20-1-1)
SPARGANIACEAE

CRB

78. Cattail

Typha latifolia Linnaeus

The slender, green, upper spike of staminate flowers withers soon after the pollen is shed leaving the compact, brown, cylindrical spike, 15–30 cm long, made up of hundreds of minute pistillate, or seed-producing, flowers. The flat linear leaves may be 2 meters long. (Two similar species in our area, *T. angustifolia* and *T. domingensis*, have slightly convex leaves.)

These rhizomatous aquatic perennials, widespread over North America, are common in the shallow water of swamps, marshes, ponds, and streams throughout Florida.

A-OB/SGE/IAB
(L 23-1-3; R 19-1-1)
TYPHACEAE

BJT

79. Everglades Palm *

Acoelorrhaphe wrightii (Griesb & Wendl.) Wendland ex Bec

Large clumps or colonies of these slender, spiny-stemmed, tropical palms often grow from a common root system and the larger trees in the colony may be 10 meters tall. The simple but palmately cleft leaves, a meter wide, are in a crowded terminal spiral.

Native to the Everglades but widely planted as an ornamental in south and central Florida.

T-3N/SRC/PAW
(L 34-8-1)
ARECACEAE

BJT

80. Coconut Palm

Cocos nucifera Linnaeus

These graceful trees can be from 3–10 meters tall. The "milk" of the large fruits is an excellent drink, and the sweet "meat" of the coconut is a nutritious food. Flowers and fruits may be on the tree at the same time.

Coconut trees, characteristic of the tropics, are naturalized in the coastal areas of south Florida and the Keys where they are also often planted as ornamentals.

T-3N/PLE/PAW
(L 34-1-1)
ARECACEAE

CRB

81. Florida Royal Palm *

Roystonea elata (Bartr.) F. Harper

The trunk, or bole, of this stately palm may be 40 meters tall and the large pinnate leaves are 3 meters or more long. The long, smooth, green sheathing leaf bases and the straight bole, that usually bulges at the base, are distinctive.

The Royal Palm is rare as a native in the rich moist hammocks of Collier, Dade and Monroe counties, but is widely planted as an ornamental in south Florida.

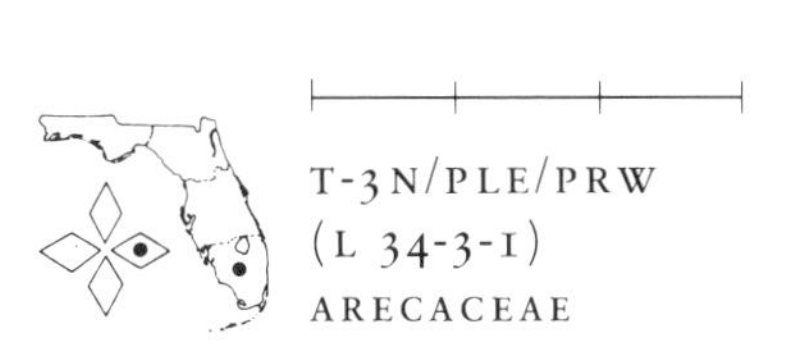

T-3N/PLE/PRW
(L 34-3-1)
ARECACEAE

BJT

82. Cabbage Palmetto

Sabal palmetto (Walt.) Loddiges ex Schultes

The state tree of Florida, this attractive palm may be 20 meters tall and is now widely used as a landscape plant. In the past the terminal buds were cut out—killing the tree—to provide "palm cabbage." The black fleshy fruits were an important food for the Indians. The trunks are used as dock pilings.

Native to coastal shell middens, hammocks and prairies more or less throughout Florida and, sparingly, along the coast to North Carolina.

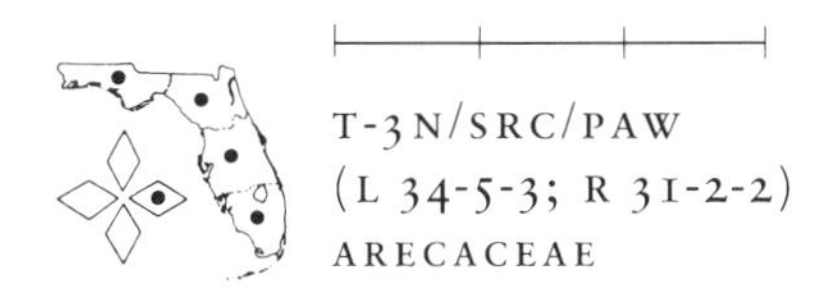

T-3N/SRC/PAW
(L 34-5-3; R 31-2-2)
ARECACEAE

DNP

83. Saw Palmetto

Serenoa repens (Bartr.) Small

The fan-shaped leaves of these shrubs may be a meter or more across, and the harsh, sawlike teeth, or spines, along the petiole of these low palms give it its common name. The black, juicy fruits, about 2 cm long, were an important Indian food. (The petioles of Blue Stem, *Sabal minor*, are smooth.)

Common on sandy prairies, dunes, and pinelands throughout Florida, Saw Palmetto also ranges along the coast to Louisiana and the Carolinas.

S-3N/SRC/PAW
(L 34-7-1; R 31-1-1)
ARECACEAE

BJT

84. Quill-leaf*

Tillandsia fasciculata Swartz

The long, slender, tapering, bluish green leaves of this conspicuous epiphyte have brown bands at the base and may be up to a meter long. The inflorescence consists of a fascicle, or cluster, of flowering spikes, each 7–15 cm long, bearing numerous colorful red to yellow or ivory bracts.

A frequent perennial on cypress, and occasionally on older pines, throughout peninsular Florida.

E-3B/SGE/ITR
(L 39-1-8)
BROMELIACEAE

BJT

85. Air Plant*

Tillandsia pruinosa Swartz

The leaves of these small, silvery white, tropical epiphytes appear to have elongate bulbous bases which are actually pseudobulbs and which usually overtop and hide the short flower stalks. The plants are only 1–3 dm tall. (The spike of the similar *T. circinata*, also found only in the Big Cypress Swamp, is longer than the leaves.)

This Bromeliad is found in Florida, at its northern limit, only on trees in the Big Cypress Swamp west of the Everglades.

E-OB/SLE/ITB
(L 39-1-10)
BROMELIACEAE

BJT

86. Wild Pine*

Tillandsia setacea Swartz

The tufts of slender leaves of these Bromeliads are 1–3 dm long and even when not in bloom form colorful clumps on tree trunks and larger branches. The spike of purple flowers is no longer than the leaves.

These common tropical epiphytes are found on trees in hammocks and swamps as far north as the Ocala area.

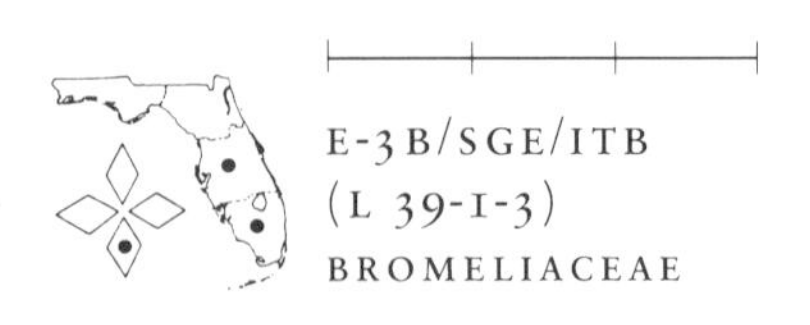

E-3B/SGE/ITB
(L 39-1-3)
BROMELIACEAE

CRB

87. Spanish Moss

Tillandsia usneoides (L.) Linnaeus

These rootless epiphytes have filiform leaves 3–5 cm long on wiry, pendulous stems that may be 1–2 meters long. The small, solitary, axillary, inconspicuous, fragrant yellow flowers produce slender brown, three-parted capsules 1.5–3 cm long.

Common on oaks, and often on other trees, throughout Florida and coastal areas of the Southeast.

E-3A/SGE/SFY
(L 39-1-1; R 37-1-1)
BROMELIACEAE

WSJ

88. Yellow-eyed Grass

Xyris fimbriata Elliott

Each of the compact, scale-covered, brown spikes, 1–2.5 cm long, produces a number of small, ephemeral, bright yellow flowers. The flat, linear basal leaves of these perennial aquatics are 3–5 dm long or about half as long as the slender scape. (*Xyris fimbriata* is the largest of some fifteen related and generally similar species in our area.)

One or more of our species of *Xyris* occurs with varying frequency in a variety of moist habitats including swamps, shallow ponds, and wet ditches from central to northern Florida and on to Mississippi and the Carolinas.

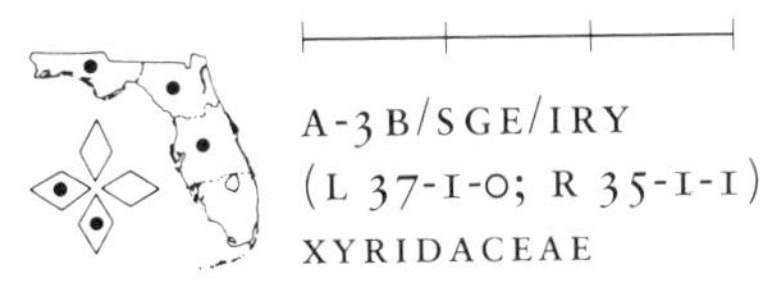

A-3B/SGE/IRY
(L 37-1-0; R 35-1-1)
XYRIDACEAE

RCS

89. Water Hyacinth

Eichhornia crassipes
(Mart.) Solms

These colonial perennials with inflated petioles form large floating mats that can completely cover lakes, ponds, and streams. The showy two-lipped, or zygomorphic, flowers are 3–5 cm across.

A prolific tropical weed now naturalized in waterways throughout our state and the frost-free coastal areas of the Southeast.

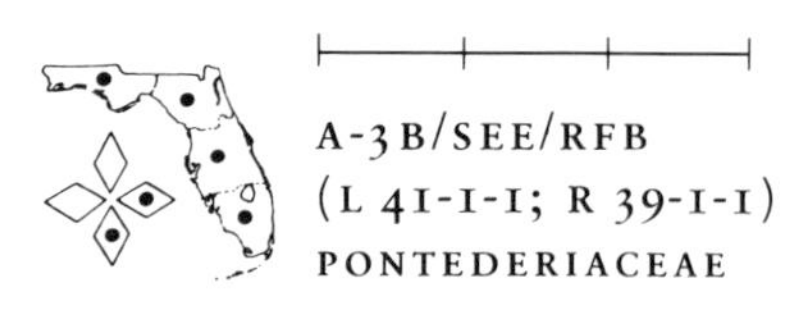

A-3B/SEE/RFB
(L 41-1-1; R 39-1-1)
PONTEDERIACEAE

DNP

90. Pickerel Weed

Pontederia cordata Linnaeus

The flower spikes of this rank aquatic are 5–15 cm long. The leaves may be nearly a meter tall and are somewhat variable in shape from cordate to lanceolate. (Accordingly, the similar *P. lanceolata*, also attributed to our area, may not be a biologically distinct species.)

This common weedy perennial, which also reproduces vegetatively, may completely fill shallow ponds, streams, marshes, and wet ditches; it occurs throughout Florida and the Southeast.

A-3B/SCE/IFB
(L 41-2-1; R 39-2-1)
PONTEDERIACEAE

BJT

91. Sawgrass

Cladium jamaicensis Crantz

This coarse rhizomatous perennial has small, cutting spines on the margins and midvein of the rigid leaves, but it is a member of the sedge family and not the grass family.

Sawgrass is common in our area and in brackish marshes from Texas to Virginia; it is the dominant plant of the "sea of grass" in the Everglades.

A-OB/SGS/PRG
(L 33-13-1; R 30-14-2)
CYPERACEAE

BJT

BJT

92. Sedge

Cyperus retrorsus Chapman

Sedges are generally wind pollinated and have minute flowers; thus many species are differentiated on small technical characters. Nonetheless, many species have attractive inflorescences and fruit clusters. Sedges usually have triangular stems that distinguish them from the round-stemmed grasses.

One or more of the various species of *Cyperus* are frequent to common in marshes, wet ditches, pond margins, and wet pinelands of each region of Florida and the eastern United States.

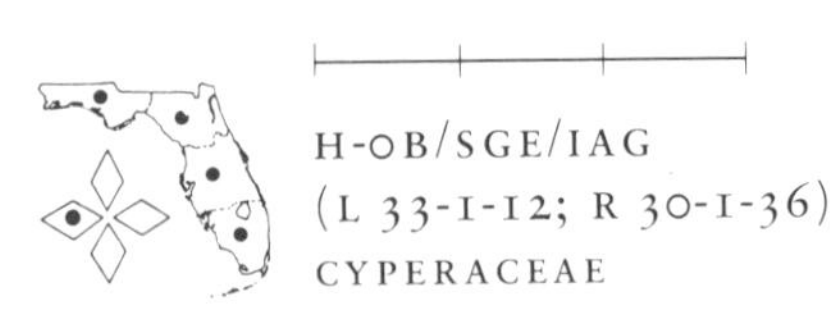

H-OB/SGE/IAG
(L 33-1-12; R 30-1-36)
CYPERACEAE

93. White Bracted Sedge

Dichromena latifolia Baldwin ex Elliott

Most sedges are wind pollinated and all have small, inconspicuous flowers. Several species, however, appear to be insect pollinated and have bright white bracts below the cluster of small flowers. These rhizomatous perennials grow to 1 meter tall. (Both *D. colorata* and the endemic *D. floridensis* have fewer than seven white bracts.)

Frequent in wet ditches, pond margins, and wet pinelands of our state and also on to Texas and Virginia along the coast.

H-OB/SGE/UAW
(L 33-10-1; R 30-4-1)
CYPERACEAE

DNP

94. Day Flower

Commelina erecta Linnaeus

These erect perennials have ephemeral flowers, about 2.5 cm across, with two large showy blue petals and one much smaller white petal. (Two related species—the common and widespread *C. diffusa* and *C. elegans*, which is only found in the Keys—both have trailing, or decumbent, stems.)

Locally frequent in open pinelands, scrub, and dry woods from south Florida to Texas and North Carolina.

H-3B/SGE/TZB
(L 40-1-2; R 38-1-2)
COMMELINACEAE

BJT

95. Pink Spiderwort

Cuthbertia graminea Small

The flowers of the Spiderworts, unlike those of the Dayflowers, are not borne within a spathe; also, all three petals are the same size. The flowers of this attractive herb are about 2 cm across, and the clustered rosette of narrow leaves is grasslike. (Two similar species, *C. rosea* and *C. ornata*, are also found in our area.)

An infrequent perennial of the sandhills and pine woods of Florida and the Atlantic Coast states north to Virginia.

H-3B/SGE/URR
(L 40-4-1; R 38-3-1)
COMMELINACEAE

CRB

96. Spiderwort

Tradescantia ohiensis Rafinesque

These rhizomatous, leafy-stemmed perennials often form large clumps with numerous stems bearing open umbels of flowers about 2 cm across. The linear leaves are 3–4 dm long.

Widespread over much of the eastern United States, Spiderworts are found in varying degrees of frequency in disturbed areas, along woodland margins, roadsides, and pond or stream margins in all parts of Florida.

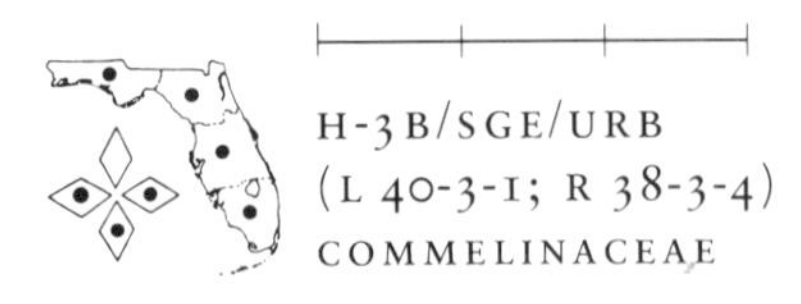

H-3B/SGE/URB
(L 40-3-1; R 38-3-4)
COMMELINACEAE

RCS

97. Hat Pins

Eriocaulon compressum Lamarck

The compact flower heads, 1–1.5 cm broad, are solitary at the end of a scape or flower stalk, up to 0.5 meters tall, from a cluster of sheathed, linear basal leaves. (The very similar *E. decangulare*, also throughout our area, has harder flower heads on scapes to 1 meter tall.)

Common in wet soil or shallow water or roadside ditches, ponds, glades, and flatwoods throughout Florida and along the coast to Mississippi and the Carolinas.

A-0B/SGE/KAW
(L 38-1-3; R 36-1-3)
ERIOCAULACEAE

WSJ

98. Bog Buttons

Lachnocaulon anceps (Walt.) Morong

Except for their smaller size, these perennial herbs are very similar in general appearance to the related but larger Hat Pins that may often be found in the same area. Bog Button flower heads are only 5–8 mm broad and the scapes only 1–5 dm tall.

Frequent on wet soils or in shallow water of roadside ditches, ponds, and prairies throughout Florida and along the coast to Texas and New Jersey.

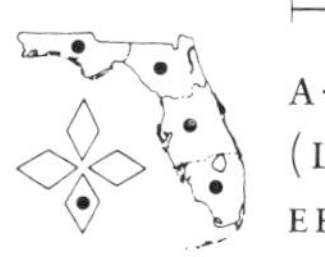

A-OB/SGE/KAW
(L 38-2-1; R 36-3-3)
ERIOCAULACEAE

CRB

99. Broom Sedge

Andropogon glomeratus (Walt.) Britton, Stern & Poggenberg

The strong feathery culms, or flower stalks, of this perennial grass (not a sedge, despite the common name) may be 1.5 meters tall and were once used to make brooms. Twelve other, somewhat similar, species of *Andropogon* also grow in our state.

Bush Bluestem, as this plant is also called, is a weed of old fields, roadsides, glades, and open pinelands throughout Florida and the Southeast.

H-OB/SGE/RAG
(L 32-54-1; R 29-87-7)
POACEAE

BJT

AHL

100. Coast Sandspur

Cenchrus incertus M. A. Curtis

The usually prostrate culms of this grass may be a meter long, and the sharp spines on the hard, 1-cm-broad fruits of this, and our six or seven other species of Sandspur, are likely known to all who have ever gone barefoot over beach dunes, sandy pinelands, or weedy lawns!

This perennial weed of beaches and dunes is common throughout Florida and much of the coastal Southeast.

H-OA/SGE/RAG
(L 32-33-5; R 29-69-4)
POACEAE

101. Beach Grass

Panicum amarum Elliott

The stout culms, or flower stalks, of this coarse, clump-forming, salt-tolerant perennial grass may be 1 cm thick at the base and up to 1 or 2 meters long. The young leaves are bluish green.

A frequent grass of beach dunes, salt marshes, and moist brackish or saline disturbed areas throughout the coastal portions of Florida and other southeastern states.

H-OB/SGE/RAG
(L 32-48-38; R 29-81-5
POACEAE

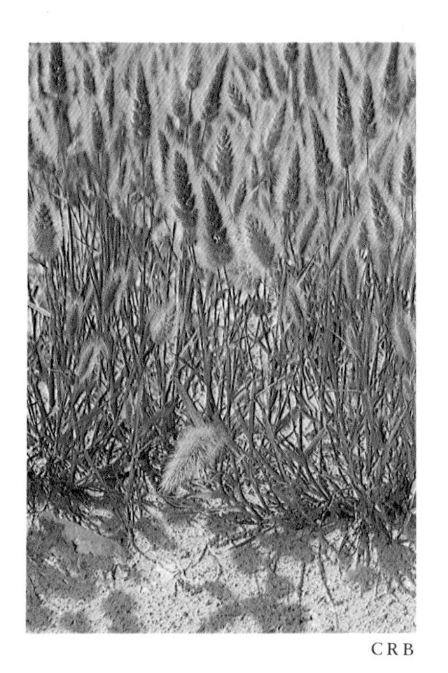

CRB

102. Rabbit Foot Grass

Polypogon monspeliensis (L.) Desfontaines

The soft, compact, spikelike panicle of this decorative, introduced annual grass is 4–8 cm long and perhaps accounts for the common name.

An occasional escape established in disturbed areas of saline sands or brackish water at scattered localities along the Florida coasts, generally north of Lake Okeechobee, west to Mississippi, and north to the Carolinas.

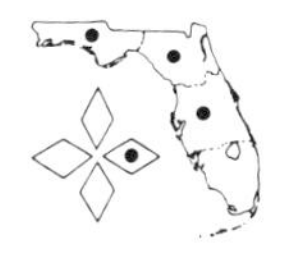

H-OB/SGE/RAG
(R 29-41-1)
POACEAE

CRB

103. Sea Oats *

Uniola paniculata Linnaeus

The stout but graceful culms of this characteristic grass of beach dunes may be 1–2 meters tall and the dense panicle of tan, often sterile, fruits or grains is 2–5 dm long. In many areas this rhizomatous perennial is planted to help stabilize sandy dunes and is protected by law.

Frequent along coastal dunes and beaches of Florida and other southeastern states.

H-OB/SGE/RAG
(L 32-19-1; R 29-10-4)
POACEAE

MIC

104. Wild Rice

Zizania aquatica Linnaeus

The profusely branched culm, or flower stalk, of *Zizania*, made conspicuous by the large reddish brown anthers of the numerous flowers, can be 3 meters tall. The slender, starchy grains of this native aquatic are considered a delicacy.

Colonies of these annuals occur occasionally in brackish or freshwater marshes of Florida and much of eastern North America.

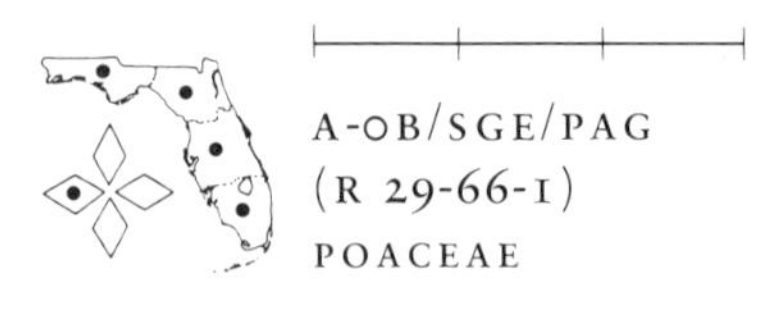

A-OB/SGE/PAG
(R 29-66-1)
POACEAE

BJT

105. Canna

Canna flaccida Salisbury

The large elliptic to lanceolate leaves of this rhizomatous perennial are 3–5 dm long  and 5–15 cm wide. The showy flowers are up to 6 cm long.

Rare or infrequent in wet ditches, marshes, and swamp margins of Florida and coastal areas to Mississippi and the Carolinas.

H-3B/SEE/IZY
(L 51-1-1; R 42-1-1)
CANNACEAE

CRB

106. Arrowroot

Thalia geniculata Linnaeus

The lanceolate leaves of *Thalia*, with blades 3–5 dm long and about half as wide, have petioles up to 1 meter long. The lax, branching inflorescence may be 2 meters tall.

Relatively frequent along streams, wet ditches, and pond margins of south and central Florida; less frequent to rare in northern Florida, and north to South Carolina.

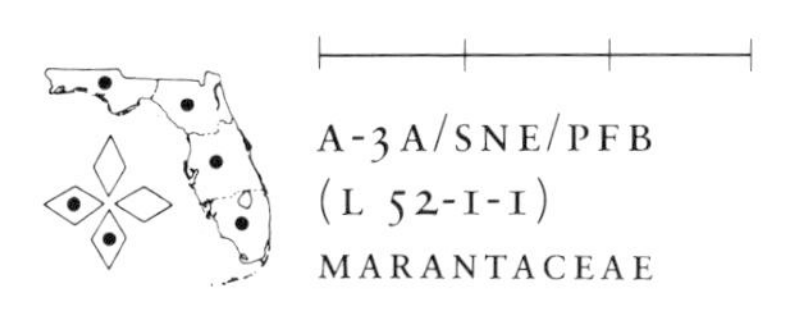

A-3A/SNE/PFB
(L 52-1-1)
MARANTACEAE

CRB

107. Star Anise

Illicium floridanum Ellis

The strongly scented flowers of this aromatic evergreen shrub have linear petals over 1 cm long. Both the flowers and fruits are star-shaped and give the plant its common name. (The related *I. parviflorum* has yellow flowers.)

Star Anise is native to the swamp margins and low hammocks of northern Florida and adjacent Gulf Coast states, but may be cultivated well outside of this range.

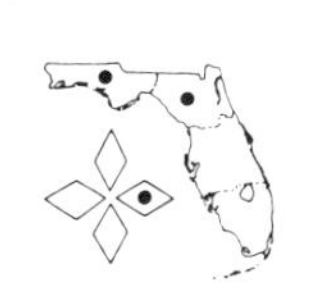

S-9A/SEE/SRB
ILLICIACEAE

AMM

108. Tulip Tree

Liriodendron tulipifera Linnaeus

A large, handsome, and commercially important tree easily recognized by its characteristic leaf shape and the brilliant orange and green tulip-shaped flowers 3–5 cm across.

A native of the eastern United States, Tulip Poplars, as they are also called, are found in rich or damp woods and along streams of northern Florida and, rarely, as far south as Orlando.

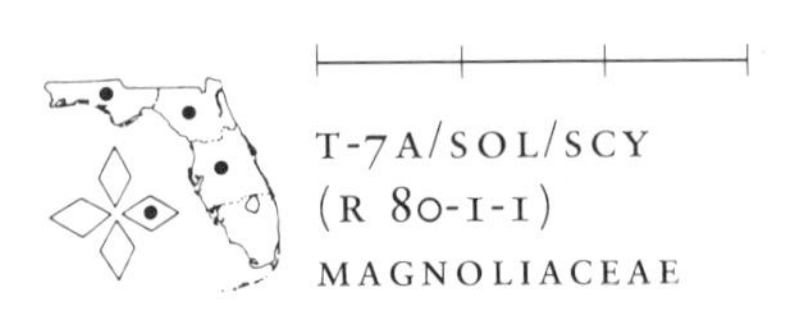

T-7A/SOL/SCY
(R 80-1-1)
MAGNOLIACEAE

MIC

109. Umbrella Tree *

Magnolia ashei Weatherby

The petals of the large, fragrant flowers may be as much as 2 dm long. The huge auriculate, or basally lobed, leaves up to 1 meter long are the largest simple leaf of any of our native trees. (The large leaves of *M. tripetala*, rare in west Florida, are not auriculate.)

Ashe's Magnolia is endemic to a few bluffs and bayheads of the Florida Panhandle. The largest known specimen of this rare tree is in the Torreya State Park.

T-7A/SBE/SRW
(R 80-2-3)
MAGNOLIACEAE

WSJ

110. Magnolia; Bull Bay

Magnolia grandiflora Linnaeus

The glossy evergreen leaves of this large tree are 10–25 cm long and reddish brown beneath. The fragrant flowers are similar in form to those of Sweet Bay but are much larger. The red seeds are eaten by birds.

These handsome trees, widely planted as ornamentals, range from the hammocks and low woods of central Florida north over much of the southeastern coastal plain.

T-7A/SEE/SRW
(R 80-2-2)
MAGNOLIACEAE

BJT

111. Sweet Bay

Magnolia virginiana Linnaeus

A medium to small tree, or a bushy stump sprout in cut over or burned areas, Sweet Bay has generally evergreen leaves, 8–15 cm long, whitish beneath and fragrant when crushed. The fragrant flowers are 3–5 cm across.

A native of the southeastern coastal plain, Sweet Bay is relatively frequent in bayheads, swamps, and wet woods more or less throughout our state.

T-7A/SEE/SRW
(L 81-1-1; R 80-2-1)
MAGNOLIACEAE

BJT

112. Pondapple; Custard Apple

Annona glabra Linnaeus

This small tree, 7–12 meters tall, often has a buttressed trunk. The flowers, about 3 cm long, are solitary on short, drooping stalks between the alternate, simple, ovate leaves. The edible fruit is 7–10 cm long.

Native to ponds, swamps, and mangrove thickets of south Florida and the Keys; rarely found along the coasts as far north as Brevard and Sarasota counties.

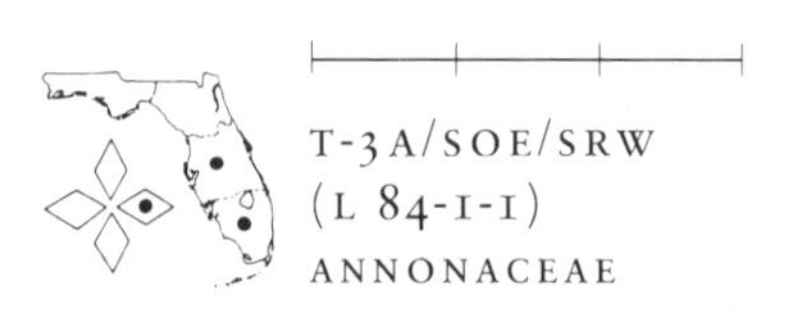

T-3A/SOE/SRW
(L 84-1-1)
ANNONACEAE

BJT

113. Pawpaw

Asimina angustifolia Rafinesque

The stiff branches of this low native shrub are usually a meter long. The fragrant, showy flowers, which appear after the leaves, are 3–8 cm long. (The narrow leathery leaves of the related *A. obovata*, which ranges into south Florida, are widest above the middle.)

A plant of the pine flatwoods, old fields, and roadsides of northern Florida and adjacent portions of Georgia and Alabama.

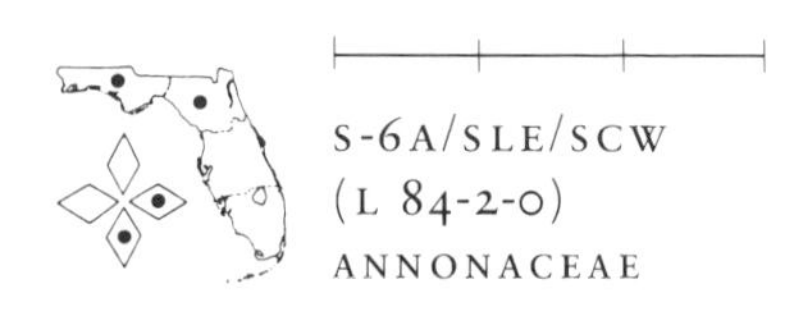

S-6A/SLE/SCW
(L 84-2-0)
ANNONACEAE

BJT

114. Flag Pawpaw

Asimina speciosa Nash

In this low, stiffly branched shrub, the 4–7 cm long flowers appear in early spring before the densely pubescent leaves. The oblong, yellow-green fruit is 5–8 cm long. (The newly emergent leaves of *A. reticulata*, endemic to peninsular Florida, are only puberulent.)

This somewhat weedy shrub is often frequent on sand ridges, in old fields and pine flatwoods from central Florida through northern Florida into Georgia.

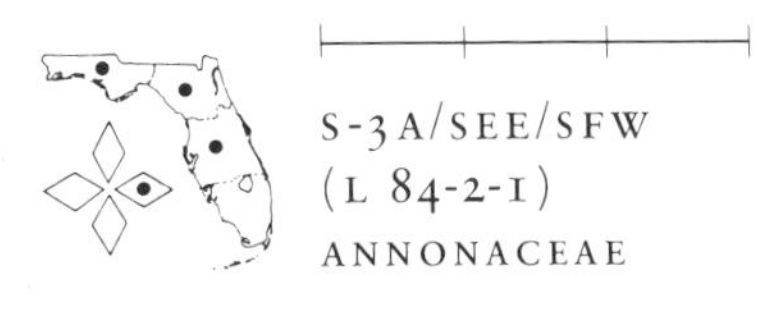

S-3A/SEE/SFW
(L 84-2-1)
ANNONACEAE

WSJ

115. Wild Ginger

Hexastylis arifolia (Michx.) Small

These low, aromatic, temperate herbs with evergreen, cordate, or heart-shaped, leaves have hidden clusters of fleshy brown or maroon flowers 1.5–3 cm long. Despite the common name, these plants are no relation to the tropical ginger from which the commercial spice ginger is produced.

Infrequent perennials of dry deciduous or pine woods from northern Florida to Alabama and Virginia.

H-3B/SCE/SCB
(R 62-3-1)
ARISTOLOCHIACEAE

RCS

116. Sweet Shrub

Calycanthus floridus Linnaeus

This native shrub, often widely planted as an ornamental, has opposite, deciduous, aromatic leaves and fragrant flowers 2.5 cm or more broad. The brown seeds are poisonous.

Sweet Betsy, or Strawberry Shrub, as this plant may also be called, is locally frequent in rich woods and woodland margins at a few localities of north Florida and, more abundantly, northward over much of the Southeast.

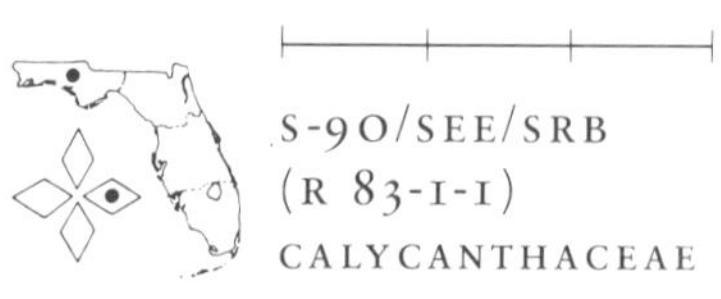

S-90/SEE/SRB
(R 83-1-1)
CALYCANTHACEAE

BJT

117. Love Vine

Cassytha filiformis Linnaeus

The somewhat fleshy, leafless, rootless, yellow green filiform stems of this tropical parasite often form thick mats on various shrub or herbaceous host plants and closely resemble the unrelated Dodder (*Cuscuta* spp.) of north temperate areas. However, the round fruits of *Cassytha*, 5–7 mm in diameter, are larger than those of *Cuscuta*.

Love Vine is frequent on various hosts in a variety of habitats from the Keys into central Florida, the northern limit of its range.

E-6N/---/RRW
(L 85-1-1)
LAURACEAE

WSJ

118. Spice Bush

Lindera benzoin (L.) Blume

These aromatic shrubs, related to the Cinnamon Tree, are deciduous and dioecious, that is, they shed their obovate 6- to 12-cm long leaves each fall and the flowers of one plant are either male or female. The brilliant red berries of the female plants are 5–8 mm in diameter.

Rare in wet woods at the southern limit of its range in the vicinity of Orlando, more frequent in low woods and along streams in northern Florida and other eastern states.

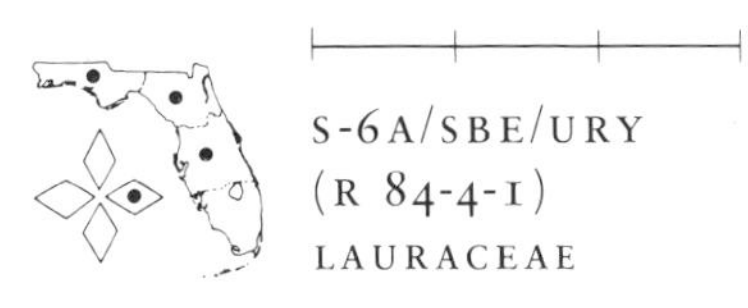

S-6A/SBE/URY
(R 84-4-1)
LAURACEAE

CRB

119. Carolina Bay

Persea borbonia (L.) Sprengel

The aromatic, evergreen elliptic leaves of this native bay, are 5–15 cm long and glabrous beneath. They have long been used either fresh or dried to flavor soups, stews, and seafood. (The leaves of *P. palustris* are tomentose beneath; the cultivated avocado, *P. americana*, from Mexico, sometimes escapes in south Florida.)

Carolina Bay is frequent in swamps and wet woods from Florida along the coastal plain to Virginia and Texas.

T-3A/SEE/UFG
(L 85-3-3; R 84-1-1)
LAURACEAE

CRB

120. Sassafras

Sassafras albidum (Nutt.) Nees

The leaves of this small deciduous, dioecious, aromatic tree are 6–20 cm long and on a single tree may be either entire, two-lobed, or three-lobed. The bark of the root is used for tea and to flavor root beer; the finely powdered dried young leaves furnish the mucilaginous "gumbo" of Creole cooking.

Sassafras is frequent to common along fencerows, old field margins, and in cut over areas of much of the Southeast; it is less frequent in Florida where it reaches its southern limit around Tampa.

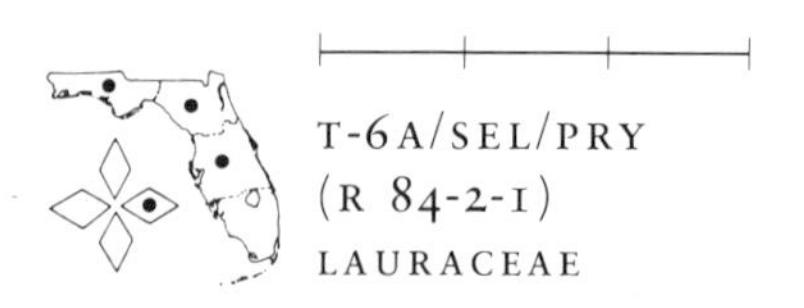

T-6A/SEL/PRY
(R 84-2-1)
LAURACEAE

BJT

121. Pale Green Peperomia

Peperomia simplex Hamilton

The flowering spike of this Peperomia can be 5–10 cm long, and the pale green elliptic leaves are 3–5 cm long.

Five species of these vinelike, often epiphytic tropical evergreen plants, some cultivated as ornamentals, are found on rotting logs or on trees in the hammocks and cypress swamps of south Florida and the West Indies; a sixth species, *P. floridana*, is endemic to south Florida.

E-OA/SOE/IAW
(L 57-1-6)
PIPERACEAE

BJT

122. Lizard's Tail

Saururus cernuus Linnaeus

The nodding or drooping tip of the 1–2 dm long raceme and the pointed cordate, or heart-shaped, leaves are characteristic of this aquatic perennial that may be 3–10 dm tall.

A rhizomatous weed of the eastern United States that is common in swamps, lake margins, small streams, and wet woodlands throughout Florida.

A-OA/SCE/RAW
(L 56-1-1; R 50-1-1)
SAURURACEAE

RCS

123. Texas Anemone*

Anemone berlandieri Pritzel

The central cylindrical "cone" of numerous individual carpels in the center of each flower is 2–3 cm long. Each carpel will produce a single wooly seed. The flowers of these low plants have no petals, but the 1–2 cm long sepals are pigmented and petal-like.

A rare perennial of open rocky woods and calcareous hammocks of Citrus, Levy, and Taylor counties in Florida and, presumably, along the Gulf to Texas and Arkansas.

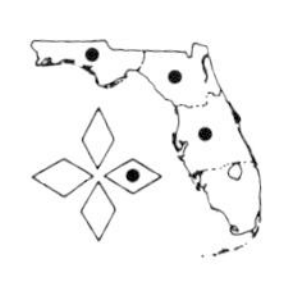

H-7A/SOC/SRW
RANUNCULACEAE

BJT

124. Columbine

Aquilegia canadensis Linnaeus

The brilliant, pendent flowers are pollinated by hummingbirds that seek the nectar at the ends of the spurs, 15–20 mm long, just the length of the hummingbird's bill. The smooth, lobed, light green, basal leaves are ternately compound.

Another rather common northern woodland perennial that is rare in our area and known only from calcareous woods of three counties—Washington, Jackson, Liberty—in west Florida.

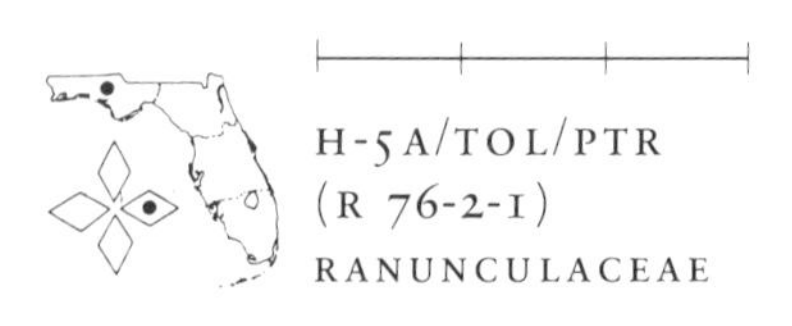

H-5A/TOL/PTR
(R 76-2-1)
RANUNCULACEAE

BJT

125. Leather Flower

Clematis crispa Linnaeus

The pale lavender, bell-shaped flower is 2.5–5 cm long and has no petals but is formed by four fleshy or leathery sepals. The pinnately compound leaves of this climbing or trailing vine have three to five linear to ovate leaflets. (The related *C. baldwinii* or Pine Hyacinth, with erect stems, is endemic to peninsular Florida.)

An occasional inhabitant of wet woods, marshes, and flood-plains more or less throughout Florida and much of the Southeast.

V-40/POL/SCB
(R 76-10-2)
RANUNCULACEAE

BJT

126. Rue Anemone

Thalictrum thalictroides (L.) Eames & Boivin

Each slender stalk of these low woodland perennials bears one to five flowers, each with five to ten white sepals that are 5–15 mm long. The ovate leaflets of the biternately compound leaves are 1–2.5 cm wide.

Windflower, as this plant is also called, is a common spring ephemeral of the more northern hardwood forests but is rare in Florida where it is known from limestone bluffs at only a few locations near Tallahassee.

H-7B/TOL/URW
(R 76-11-1)
RANUNCULACEAE

RCS

127. Prickly Poppy

Argemone albiflora Hornemann

A spiny annual with white milky sap, or latex, and conspicuous flowers 8–12 cm or more across.

An infrequent native weed along roadsides and in waste places in central and northern Florida, west to Texas and north to the Carolinas.

H-7A/SNC/URW
(R 85-2-1)
PAPAVERACEAE

WSJ

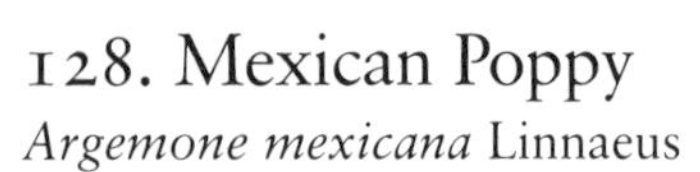

128. Mexican Poppy

Argemone mexicana Linnaeus

The sunny flowers of the Prickly Poppy (as this species is also sometimes called) are 5–10 cm across, and the mottled spiny leaves are 1–2 dm long; the sap is yellow.

This annual from Mexico is now naturalized along roadsides and in cultivated fields at scattered localities in Florida and several other southern states.

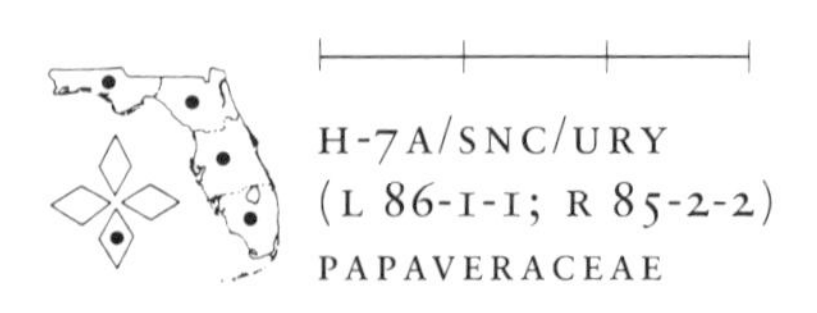

H-7A/SNC/URY
(L 86-1-1; R 85-2-2)
PAPAVERACEAE

WSJ

129. Watershield

Brasenia schreberi Gmelin

The inconspicuous maroon flowers of these aquatics are only about 1 cm across but the 4–10 cm long, shield-shaped, evergreen floating leaves (which have a thick gelatinous layer on the underside) are a conspicuous element of the aquatic flora in some ponds, lakes, and slow streams.

A cosmopolitan species found at scattered localities throughout Florida.

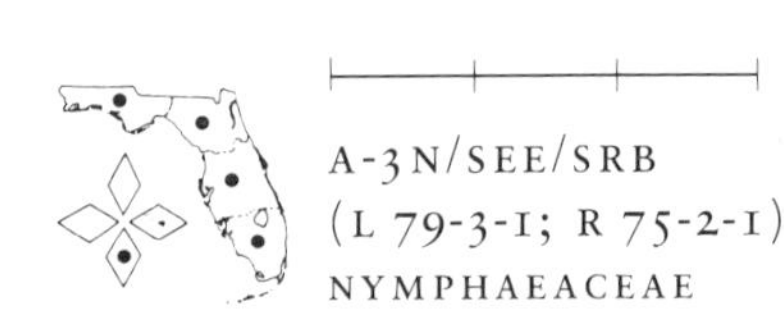

A-3N/SEE/SRB
(L 79-3-1; R 75-2-1)
NYMPHAEACEAE

RKG

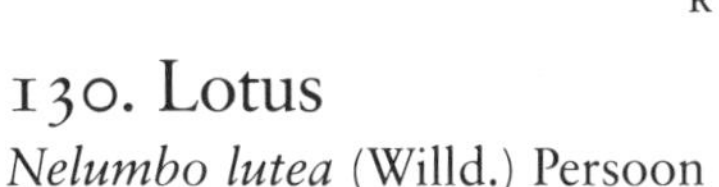

130. Lotus

Nelumbo lutea (Willd.) Persoon

The spectacular flowers of this native perennial are often 2–3 dm across and stand as much as 1 meter above the water surface with the round, light green leaves that can be 3–6 dm in diameter. (The cultivated *N. nucifera*, with pink flowers, is not native but may occasionally escape and become established.)

Infrequent in pools and quiet waters at widely scattered localities throughout Florida and much of the Southeast.

A-7N/SRE/SCY
(L 79-2-1; R 74-1-1)
NYMPHAEACEAE

BJT

131. Cow Lily; Spatter Dock

Nuphar luteum (L.) Sibthorp & Smith

The leaves of these coarse, rhizomatous perennials are quite variable and may be wide or rather narrow. The fleshy flower, 2–5 cm across, has a prominent lobed stigma.

Native to the ponds, lakes, and rivers of the lower elevations of the Southeast, these aquatics are relatively frequent throughout Florida.

A-7N/SRE/SCY
(L 79-4-1; R 73-1-1)
NYMPHAEACEAE

WSJ

132. Banana Water Lily

Nymphaea mexicana Zuccarini

The common name of this introduced water lily refers to its thick tuberlike, stoloniferous (or creeping) rhizome by which these plants can reproduce vegetatively. The emergent flowers, usually 1 dm above the floating leaves, are about 1 dm across. (The related *N. elegans*, also from Mexico, has emergent blue flowers.)

Established in lakes, ponds, slow streams, and sloughs at scattered localities throughout Florida and along the coast to Louisiana and the Carolinas.

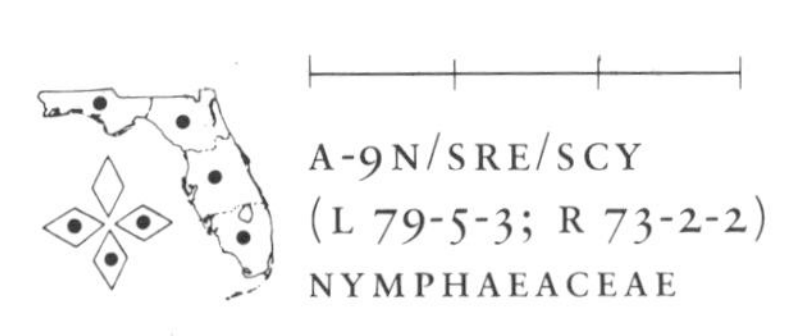

A-9N/SRE/SCY
(L 79-5-3; R 73-2-2)
NYMPHAEACEAE

DNP

133. Water Lily

Nymphaea odorata Aiton

The fragrant flowers of this rhizomatous, native aquatic are 8–15 cm across when open, but they close at night or on very cloudy days. The large floating leaves, to 3 dm across, are green above and reddish beneath.

A frequent perennial of the ponds, lakes, and ditches of Florida and much of the Southeast.

A-9N/SRE/SCW
(L 79-5-2; R 73-2-1)
NYMPHAEACEAE

BJT

134. Trumpets *

Sarracenia flava Linnaeus

Insects are trapped and digested in the long, slender, erect, hollow leaves and provide the plant with some of its mineral requirements. The leaves, often marked with red, may be 3–10 dm tall. The flowers of these interesting perennials have a strong musky scent.

A native of bogs and savannas of the southeast coastal plain that is becoming quite rare; found in northern Florida only in a few localities.

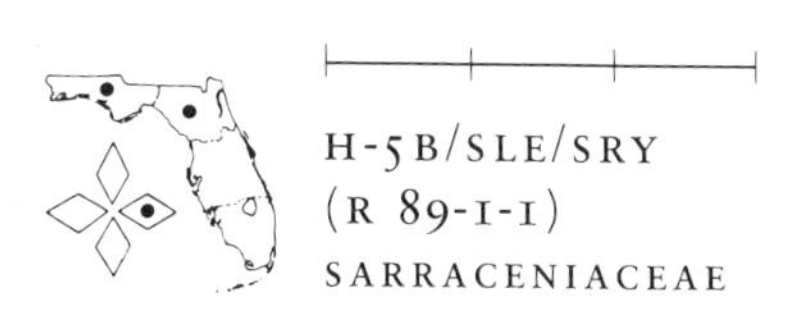

H-5B/SLE/SRY
(R 89-1-1)
SARRACENIACEAE

CRB

135. White Top Pitcher Plant *

Sarracenia leucophylla Rafinesque

The hollow, erect, white-topped leaves of this *Sarracenia*, 4–8 cm across at the top of the open "pitcher," were mistaken for flowers by early botanists, who thought this native perennial to be related to the Jack-in-the-Pulpit.

Infrequent to rare in a few bogs and low open areas of west Florida and adjacent areas of other states.

H-5B/SLE/SRR
SARRACENIACEAE

BJT

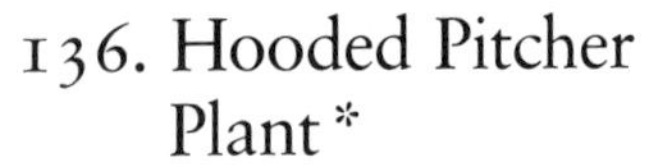

136. Hooded Pitcher Plant *

Sarracenia minor Walter

The spotted, strongly hooded, hollow, erect leaves of these plants are usually 1.5–3 dm tall and about the same height as the stalked, odorless flowers.

A native of the bogs and savannas of the Atlantic coastal plain to North Carolina, these perennials are infrequent to rare in all sections of Florida.

H-5B/SLE/SRY
(R 89-1-3)
SARRACENIACEAE

RKG

BJT

137. Sweet Pitcher Plant *

Sarracenia rubra Walter

The fragrant flowers of the Sweet Pitcher Plant are 2–3 cm broad and are on slender stems often twice as tall as the hollow, erect leaves. (The leaves of two other red-flowered species, *S. psittacina* and *S. purpurea*, inset, are decumbent.)

Rare, and fortunately rather inconspicuous, along the margins of bogs and low woods on the coastal plain from Mississippi to the Carolinas; known in Florida from a few localities in the two or three westernmost counties.

H-5B/SLE/SRR
(R 89-1-2)
SARRACENIACEAE

DNP

138. Loblolly Bay

Gordonia lasianthus (L.) Ellis

The thick, glossy evergreen leaves of this shrub or small tree are often ragged where chewed by insects. The Camellia-like flowers are 4–6 cm across.

Infrequent in bay forests and wet thickets from Lake Okeechobee north along the coast to the Carolinas and west to Louisiana.

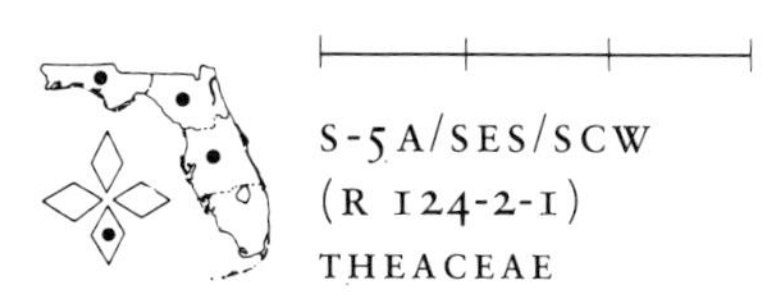

S-5A/SES/SCW
(R 124-2-1)
THEACEAE

CRB

139. Possum Haw

Ilex decidua Walter

A shrub or small tree with narrow deciduous leaves, 3–8 cm long, with shallowly toothed or serrate margins. The red or yellow fruits on the female plants are 4–7 mm in diameter.

Possum Haw is native to the southeastern United States and is frequent throughout Florida in river bottoms and low thickets, especially along the Suwannee and Aucilla rivers of north Florida.

S-5A/SES/URW
(L 112-1-7; R 112-1-4)
AQUIFOLIACEAE

WSJ

140. Gallberry

Ilex glabra (L.) Gray

This often weedy, rhizomatous shrub, 1–2 meters tall, is our smallest holly. It is easily recognized by its black fruits and the small evergreen leaves, 2–5 cm long, that are widest above the middle and often have a few low, blunt teeth near the tip.

Inkberry, as this plant may also be called, is common in low, open sandy areas and savannas throughout Florida and along the coast to Louisiana and Nova Scotia.

S-7A/SBS/URW
(L 112-1-3; R 112-1-9)
AQUIFOLIACEAE

BJT

CRB

141. Myrtle-leaf Holly

Ilex myrtifolia Walter

This shrub or small tree has leathery, or coriaceous, evergreen, elliptic leaves 5 – 10 cm long with usually smooth margins. The attractive red to yellow fruits on the female trees are 6 – 8 mm in diameter. (This species is sometimes considerd a variety of *I. cassine*, which has wider leaves.)

A tree of swamps, lake margins, andd low sandy areas throughout Florida and along the coast to Louisiana and Virginia.

S-4A/SEE/URW
(L 112-1-0; R 112-1-2)
AQUIFOLIACEAE

WSJ

BJT

142. Yaupon

Ilex vomitoria Aiton

The small, thick evergreen leaves of this large shrub or small tree contain caffeine and, when dried, were used by the Indians to brew a ceremonial "Black Drink." The elliptic leaves are 1–5 cm long; the "berries" are 4–6 mm in diameter.

Widely planted as an ornamental or as a hedge plant, Yaupon is native to the maritime forests and forest margins from central Florida to Louisiana and Virginia.

S-4A/SNS/URW
(L 112-1-1; R 112-1-3)
AQUIFOLIACEAE

143. Sweet Pepperbush

Clethra alnifolia Linnaeus

The alternate, oblanceolate to elliptic leaves of this attractive shrub are 4–10 cm long and may be either glabrous or pubescent beneath. The fragrant flowers are borne in compact racemes 1–2 dm long.

A shrub of the coastal plain from Virginia to Mississippi, Sweet Pepperbush occurs in the pocosins, bays, and pine barrens of northern Florida.

S-5A/SES/RRW
(R 143-1-2)
CLETHRACEAE

CRB

144. Black Ti-ti

Cliftonia monophylla (Lam.) Britton ex Sargent

The elliptic evergreen leaves of this shrub or small tree are 3–6 cm long, or about half the length of the compact raceme of fragrant flowers.

Black Ti-ti grows in acid swamps, bogs, and wet ditches primarily of west Florida and in only a few counties of other states along the Gulf.

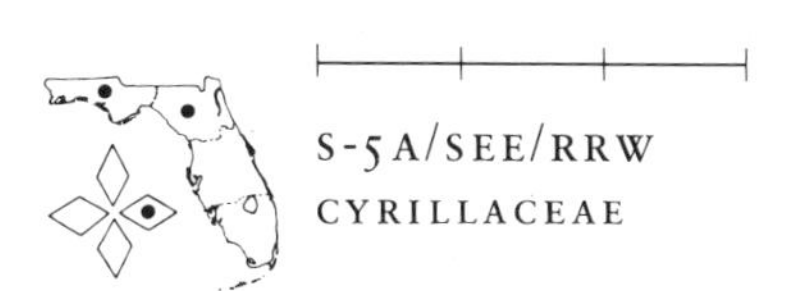

S-5A/SEE/RRW
CYRILLACEAE

BJT

145. Ti-ti

Cyrilla racemiflora Linnaeus

This semi-evergreen shrub or small tree with attractive lustrous, alternate, obovate leaves 5–10 cm long bears numerous showy, subterminal racemes 8–15 cm long.

Leatherwood, as this plant is also known, is frequent in swamps, low woods, and stream banks more or less throughout Florida, and on the coastal plain from Texas to the Carolinas.

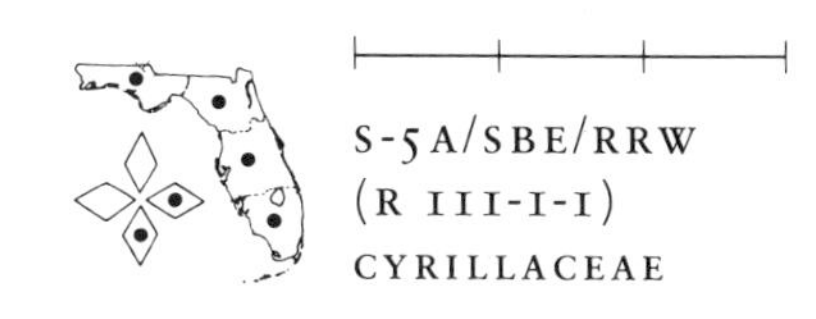

S-5A/SBE/RRW
(R III-I-I)
CYRILLACEAE

BJT

146. St. John's Wort

Hypericum cistifolium Lamarck

The conspicuous, branched, terminal inflorescence, or flower cluster, and the shiny, hollylike leaves, 2–8 cm long, help identify this small shrub. (In *H. galioides*, another of the sixteen or so species of *Hypericum* in Florida, there are leaflike bracts mixed with the flowers in the inflorescence.)

A frequent plant of wet pinelands and stream margins throughout Florida and other coastal states of the Southeast.

S-50/SEE/URY
(L 123-1-6; R 126-1-17)
CLUSIACEAE

WSJ

147. Pineweed

Hypericum gentianoides (L.) Britton, Stern & Poggenberg

These slender, much-branched annual herbs are 2–5 dm tall and have small, almost scalelike, leaves. The flowers are 4–8 mm across.

A common weed of fallow sandy fields, roadsides, and waste places from south Florida to Texas and Ontario.

H-50/SLE/SRY
(L 123-1-5; R 126-1-18)
CLUSIACEAE

DNP

BJT

148. Hypericum

Hypericum myrtifolium Lamarck

The orange yellow flowers of this evergreen shrub have obovate petals and are 2–3 cm across. The clasping elliptic leaves are 1–3 cm long.

A frequent shrub found in wet pinelands and along the margins of lakes and ponds throughout Florida, and along the Gulf to Mississippi.

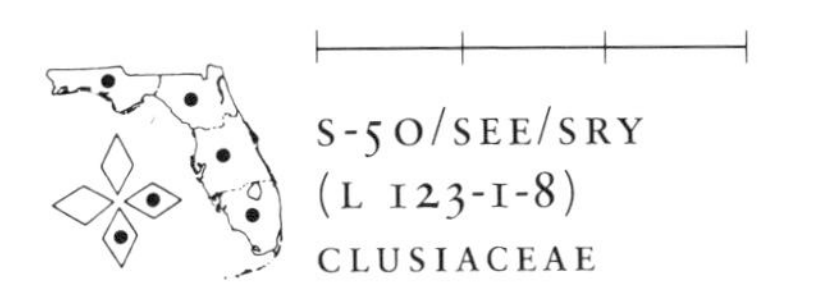

S-5 O/SEE/SRY
(L 123-1-8)
CLUSIACEAE

149. St. John's Wort

Hypericum reductum P. Adams

The herbaceous or semiwoody stems of this St. John's Wort are decumbent or spreading from the base rather than strictly erect, and the plant thus forms a low mat. The clear yellow flowers are 1–2 cm broad.

A frequent plant of sandy woods, scrub, and coastal dunes from south Florida into Alabama and the Carolinas.

S-5 O/SLE/URY
(L 123-1-10; R 126-1-11)
CLUSIACEAE

BJT

150. Tarflower

Befaria racemosa Ventenat

The white or pinkish petals of the conspicuous flowers of Tarflower are 2–3 cm long and are sticky, thus the common name. The ovate evergreen leaves of these coarsely pubescent shrubs are 2–7 cm long.

Endemic to the pine flatwoods and scrub of peninsular and north Florida and southern Georgia.

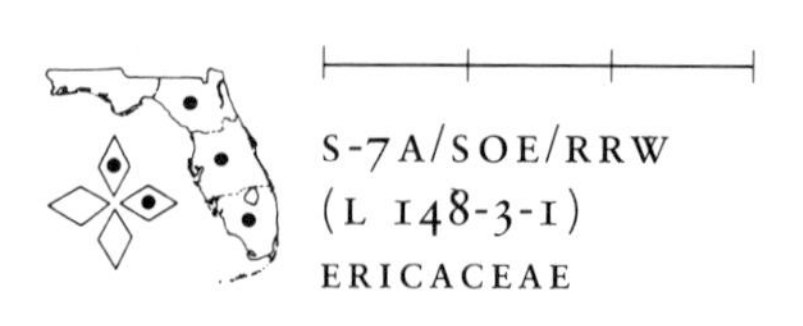

S-7A/SOE/RRW
(L 148-3-1)
ERICACEAE

WSJ

151. Dwarf Huckleberry

Gaylussacia dumosa (Andrews) Torrey & Gray

These rhizomatous shrubs are similar to the Blueberries, but their fruits each have only ten relatively large seeds; the elliptic, deciduous or semievergreen leaves, 1.5–6 cm long, have small resinous glands beneath. (*G. frondosa*, also in our area, has deciduous floral bracts.)

A southeastern plant of low pinelands, scrub, and the margins of acid ponds throughout our state but rare in south Florida.

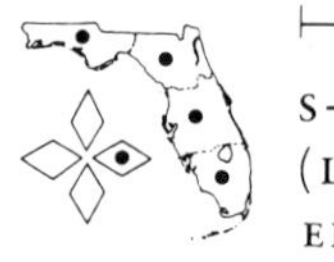

S-5A/SEE/RCW
(L 148-2-1; R 145-18-2)
ERICACEAE

BJT

DNP

152. Hairy Laurel

Kalmia hirsuta Walter

This low, evergreen shrub has hirsute or densely pubescent twigs and leaves. The small flowers, about 1.5 cm across, are solitary, or two or three, in the axils of the new leaves. (The related Mountain Laurel of the Appalachians, *K. latifolia*, is rare in northern Florida and is on the protected list for our state.)

Infrequent in pine savannas, flatwoods, and moist sandhill areas of northern Florida and on to Mississippi and Virginia.

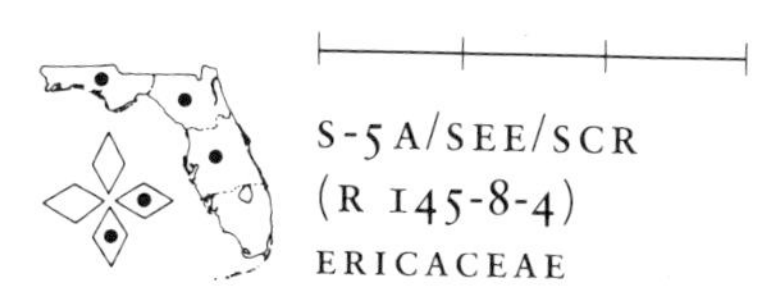

S-5A/SEE/SCR
(R 145-8-4)
ERICACEAE

BJT

WSJ

153. Fetterbush
Lyonia lucida (Lam.) K. Koch

The glossy, leathery, elliptic evergreen leaves of these shrubs are 3–9 cm long. The urn-shaped flowers, 6–8 mm long, may be pale to dark pink. (The flowers of the deciduous *L. mariana*, found in central and northern Florida, are white.)

Fetterbush is common in open pinelands throughout Florida and much of the southeastern coastal plain.

S-5A/SEE/RCR
(L 148-4-1; R 145-11-3)
ERICACEAE

154. Pine-sap *
Monotropa hypopithys Linnaeus

These low, colorful saprophytes, which live on decaying wood buried in the soil, are 1–2 dm tall and differ from Indian Pipes both in the brilliant coloration and the several flowers borne on each leafless stem.

A very rare plant of dry sandy soil at one locality in central Florida that is more common northward.

E-5N/---/RFR
(R 145-3-2)
ERICACEAE

BJT

155. Indian Pipes

Monotropa uniflora Linnaeus

These low, pale, fleshy, saprophytic perennials are sometimes mistaken for fungi but are closely related to other members of the Heath family by the structure of the solitary, nodding flowers, which are about 1.5 cm long. The plants turn black after fruiting.

A widespread but more northern species rarely found in rich deciduous woods and scrub of central and northern Florida.

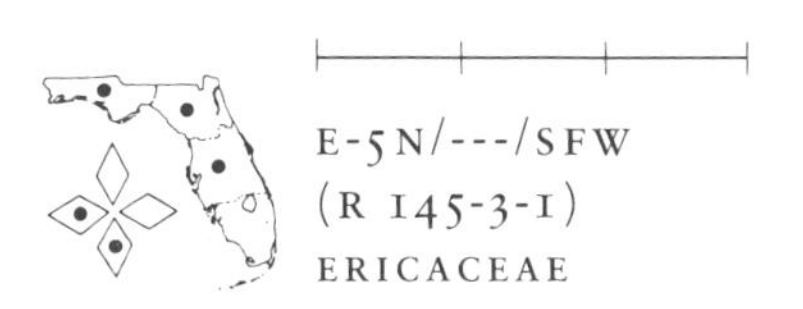

CRB

156. Sourwood

Oxydendrum arboreum (L.) de Candolle

The numerous small, urn-shaped flowers, each only 5–8 mm long, are borne in compact flat sprays. The alternate, lanceolate deciduous leaves of this tree are 1–2 dm long, finely serrate, and have a sour taste.

A more northern tree that is infrequent on the well-drained slopes and stream banks of west Florida.

BJT

157. Wild Azalea

Rhododendron canescens
(Michx.) Sweet

The slightly fragrant pink flowers, 2–3 cm across, usually appear on these large shrubs before the thin elliptic leaves.

A native of acid sandy soils along stream banks and swamp margins of northern Florida and the southeast coastal plain.

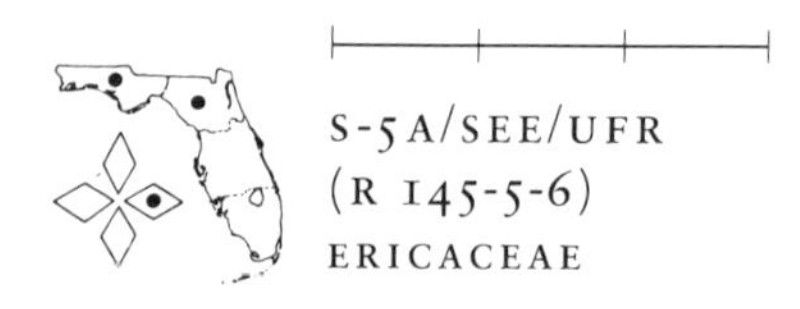

S-5A/SEE/UFR
(R 145-5-6)
ERICACEAE

SWL

158. Chapman's Rhododendron*

Rhododendron chapmanii Gray

The evergreen leaves of this rare shrub are 3-4 cm long and have many small, yellow glandular dots, especially on the lower surface. The flowers are about 3 cm long.

This endemic shrub occurs in pine flatwoods and on the borders of a few Ti-ti swamps in only four counties of northern Florida.

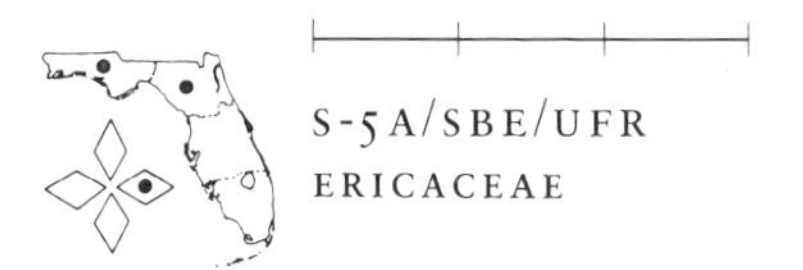

S-5A/SBE/UFR
ERICACEAE

BJT

159. Shiny Blueberry

Vaccinium myrsinites Lamarck

The small, glossy evergreen leaves of this low, erect, rhizomatous shrub are only 1–3 cm long and have stalked glands beneath. The pink to white urceolate, or urn-shaped, corolla is 6–8 mm long. (The leaves of *V. darrowii*, another of our Blueberries, are not glandular pubescent beneath.)

This is a common shrub of pine flatwoods and sandhills throughout Florida and adjacent areas of Georgia and Alabama.

S-5A/SEE/RCW
(L 148-1-1; R 145-19-1)
ERICACEAE

WSJ

160. Deerberry

Vaccinium stamineum Linnaeus

The small leafy bracts mixed with the flowers or fruits help identify this large shrub, which may be 1–4 meters tall and is often rather fully branched. The flower "buds" of Deerberry are open from their formation. The bitter green fruits are about 1 cm in diameter.

These shrubs are frequent in hammocks and flatwoods of central and northern Florida and dry woodlands of much of the Southeast.

S-5A/SEE/RCW
(R 145-19-2)
ERICACEAE

BJT

161. Low Bush Blueberry

Vaccinium tenellum Aiton

This low, erect, deciduous, rhizomatous shrub, only 1–4 dm tall, has narrowly urceolate, or urn-shaped, flowers 5–8 mm long. (The related and more frequent *V. elliottii*, which is not rhizomatous, grows to 1–3 meters tall.)

Infrequent in Florida along woodland borders only in the Gainesville area; more frequent in open coastal plain woodlands north to Virginia and west to Mississippi.

S-5A/SEE/UCW
(R 145-19-4)
ERICACEAE

CRB

162. Rosemary

Ceratiola ericoides Michaux

The numerous stiffly erect branches of this shrub, which may be 2–15 dm tall, have four rows of short, slender, needlelike leaves with small brown or yellowish flower clusters in the leaf axils.

Rosemary is frequent on open sand ridges, old dunes, and in pine scrub of central and northern Florida and along the coasts to South Carolina and Mississippi.

S-OA/SLE/UAB
(L 110-1-1; R 144-1-1)
EMPETRACEAE

163. Wild Sapodilla

Manilkara bahamensis (Baker) Lamarck & Meeuse

The round brown fruits, about 3 cm in diameter, and the elliptic evergreen leaves, which are often slightly notched at the end, help identify this tropical shrub or small tree as Wild Dilly. The pale yellow flowers are about 2 cm across. (The leaves of the naturalized, tropical *M. zapoda*, from which we get chicle for chewing gum, are not notched at the end.)

Wild Dilly is restricted to the hammocks of south Florida and the Keys, but also grows in the Bahamas and West Indies.

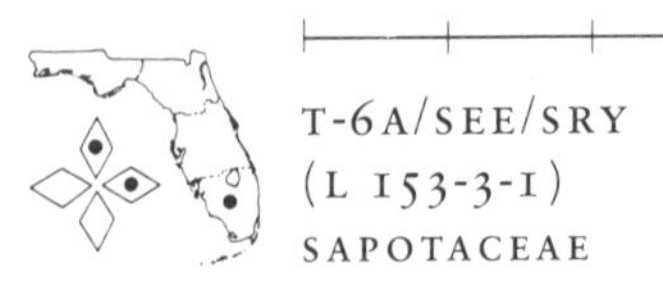

BJT

164. Horse-sugar

Symplocos tinctoria (L.) L'Heritier

A deciduous or weakly evergreen shrub or small tree, sometimes cultivated, that usually flowers before the leaves appear. The compact clusters of small, fragrant, cream-colored flowers are borne in profusion along the branches of the previous year's growth.

Primarily of southeastern distribution, Horse-sugar, or Sweetleaf, grows in alluvial woods and along streams of central and northern Florida.

T-5A/SEE/KRY
(R 151-1-1)
SYMPLOCACEAE

WSJ

165. Snowbell

Styrax grandifolia Aiton

The deciduous leaves of this shrub or small tree are stellate pubescent beneath, 5–15 cm long, and almost as wide. The fragrant waxy flowers are 2–3 cm broad. (The leaves of *S. americana*, which ranges into central Florida, are smaller, narrower, and glabrous beneath.)

Somewhat infrequent on bluffs, floodplains, and in hammocks of northern Florida and other southeastern states.

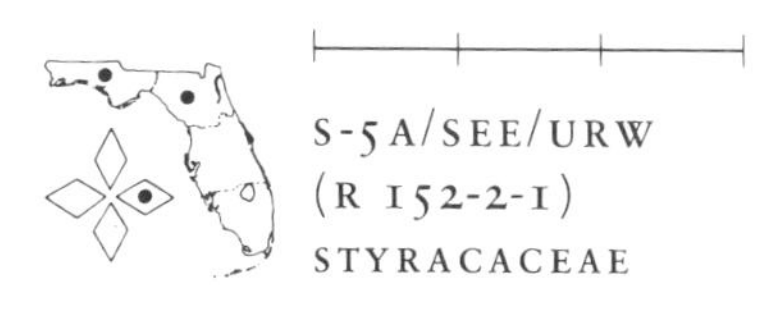

S-5A/SEE/URW
(R 152-2-1)
STYRACACEAE

BJT

166. Marlberry

Ardisia escallonioides
Schlechtendahl & Chamisso

The dense terminal panicles of small white flowers are 10–15 cm long, as are the alternate, narrowly ovate leaves of this tropical shrub or small tree.

Marlberry grows in hammocks and pinelands of south Florida and in central Florida near the coasts; it is now naturalized in the Tallahassee area. It also grows in Mexico and the West Indies.

S-5A/SEE/PRW
(L 150-2-1)
MYRSINACEAE

WSJ

167. Pimpernel

Anagallis arvensis Linnaeus

The flowers of this sprawling winter annual are about 1 cm across and are solitary on long pedicels in the axils of the alternate leaves. (The flowers of the related and more widespread False Pimpernel, *Centunculus minimus*, are white, almost sessile and have only four petals.)

This European weed of fallow fields, roadsides, and waste places is a rare introduction to Florida reported only from Orange County; it is more frequent in other southeastern states.

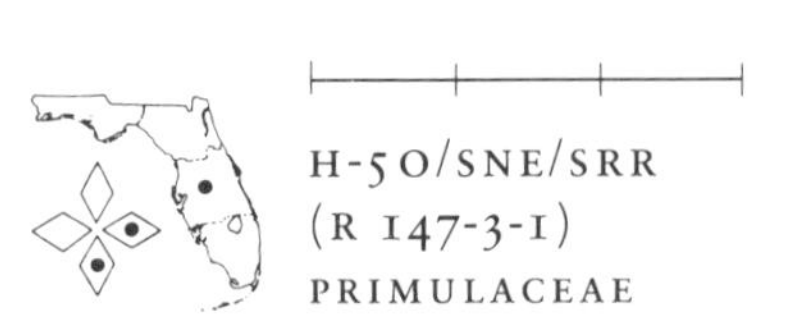

H-5O/SNE/SRR
(R 147-3-1)
PRIMULACEAE

BJT

168. Florida Violet

Viola floridana Brainerd

Florida Violet is a low rhizomatous perennial with cordate leaves 4–5 cm long and zygomorphic flowers, which means they are bilaterally rather than radially symmetrical. (The very similar *V. affinis* and *V. septemloba*, with lobed leaves, also have purple flowers and are frequent in Florida.)

This species, one of our most frequent purple violets, is found in open woods and clearings throughout Florida and on into Mississippi and the Carolinas.

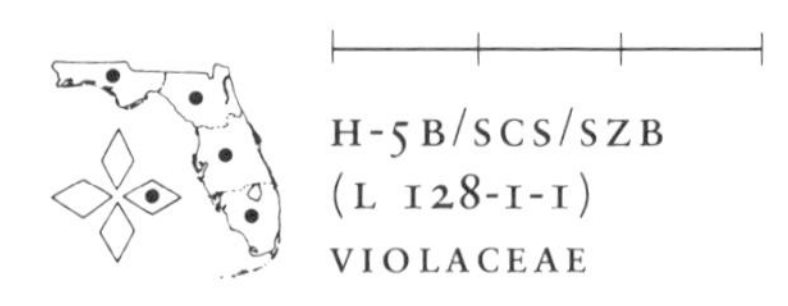

H-5B/SCS/SZB
(L 128-1-1)
VIOLACEAE

WSJ

169. Halberd Leaf Violet*

Viola hastata Michaux

Both the yellow flower and the silver gray markings between the leaf veins help identify this woodland perennial. Each of the rhizomatous plants produces only two or three leaves each year.

The Halberd Leaf Violet is quite rare in Florida, where it has been reported from only a single county of the Panhandle; however, it is more frequent northward in the deciduous woods of other southeastern states.

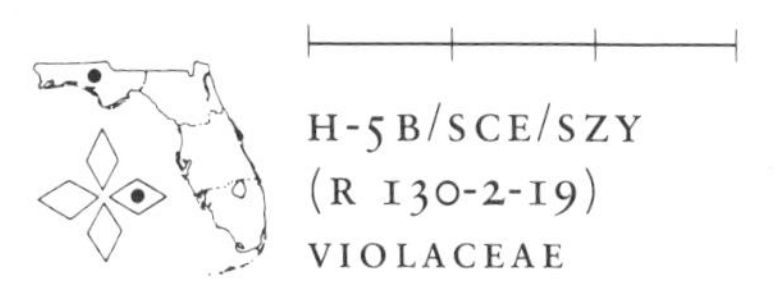

H-5B/SCE/SZY
(R 130-2-19)
VIOLACEAE

BJT

170. Long-leaf Violet

Viola lanceolata Linnaeus

The actual width of the lanceolate leaves of this small perennial violet may vary considerably from plant to plant. The zygomorphic flowers are 1–2 cm across. (The leaves of *V. primulifolia*, also with white flowers, are ovate.)

A common violet of moist, sandy, open areas throughout Florida and much of the Southeast.

H-5B/SNS/SZW
(L 128-1-2; R 130-2-16)
VIOLACEAE

RCS

171. Field Pansy

Viola rafinesquii Greene

Although the small pansylike flowers, only 1 cm across, may be white to pale blue, the erect or spreading leafy stem aids in the recognition of the Field Pansy.

This attractive little weedy annual often forms dense but ephemeral and harmless populations in lawns, clearings, old fields, and along roadsides in a few counties of west Florida and over much of the Southeast.

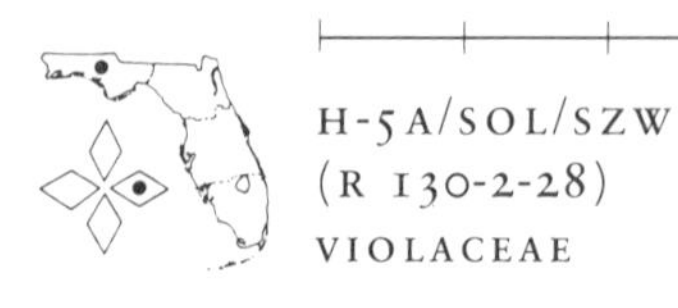

H-5A/SOL/SZW
(R 130-2-28)
VIOLACEAE

CRB

172. Annatto

Bixa orellana Linnaeus

This rather dense, rounded, shrub or small tree, with widely ovate, acuminate leaves 10–20 cm long, has clusters of pink flowers 5 cm across followed by reddish brown, spiny, ovoid capsules 3–5 cm long. The orange to red pulp around the seeds is the annatto used in cooking and as a dye.

Annatto, also called Lipstick Tree, is cultivated and apparently naturalized locally in south Florida.

S-5A/SOE/PRR
(L 126-1-1)
BIXACEAE

BJT

173. Piriqueta

Piriqueta caroliniana (Walt.) Urban

These colorful, perennial herbs, with nearly glabrous to densely stellate pubescent stems 1.5–4 dm tall and variable, pubescent leaves, often reproduce vegetatively by root sprouts and form large colonies. The flowers are 2–4 cm across.

Frequent in open, wet to dry pinelands, clearings, and along woodland margins in all sections of Florida and north on the coastal plain to the Carolinas.

H-5A/SLD/RRY
(L 130-2-1; R 125-1-1)
TURNERACEAE

RCS

174. Passion Flower

Passiflora incarnata Linnaeus

The conspicuous, intricate flowers of Apricot Vine or Maypop, are 5–8 cm in diameter and the edible ovoid fruits are 5–10 cm long. The palmately three-lobed leaves, 5–15 cm long, are finely serrate.

These trailing or climbing perennial vines are found in waste areas, old fields, and along roadsides at scattered localities throughout Florida and much of the central and southern United States.

V-5A/SEL/SRB
(L 131-1-1; R 131-1-1)
PASSIFLORACEAE

BJT

175. Yellow Maypop

Passiflora lutea Linnaeus

The wide, shallowly lobed leaves with entire margins and the small yellow green flowers 1–2 cm in diameter are characteristic of this relatively inconspicuous perennial vine. (Two other species with greenish white flowers and smooth leaf margins, the tomentose *P. multiflora* and *P. suberosa*, which has corky winged stems, occur in south Florida.)

Infrequent to rare in peninsular Florida, somewhat more frequent but widely scattered in thickets, waste areas, and open deciduous woodlands of northern Florida and much of the eastern United States.

V-5A/SOL/SRY
(R 131-1-3)
PASSIFLORACEAE

CRB

176. Papaya

Carica papaya Linnaeus

The soft, unbranched stem or trunk of this small dioecious tree is usually less than 5 meters tall and bears a crown of large, long-petioled, palmately lobed leaves 3–6 dm wide. The fleshy edible fruits are 1.5–3 dm long. Papain, used medicinally and as a meat tenderizer, is obtained from the milky sap of the leaves and green fruit.

This cultivated tropical introduction has become naturalized as seeds are distributed outside of the areas of cultivation in south and central Florida.

T-5N/SOL/PRW
(L 132-1-1)
CARICACEAE

WSJ

177. Creeping Cucumber

Melothria pendula Linnaeus

The small, pendulous, melonlike fruits of Creeping Cucumber are only 1–3 cm long; the lobed, ovate to reniform leaves are usually 2–8 cm wide, but can be quite variable.

This inconspicuous perennial vine often forms small mats on sandy roadsides, in low woods, and along the borders of marshes of Florida and other coastal states from Mexico to the Carolinas.

V-5A/SOL/SRY
(L 176-1-1; R 177-6-1)
CUCURBITACEAE

CRB

178. Wild Balsam Apple

Momordica charantia Linnaeus

These annual to perennial trailing or climbing vines, with alternate, deeply palmately lobed leaves 4–10 cm wide, have light yellow, monoecious flowers 1.5–2.5 cm across. However, the colorful, toxic, ovoid fruits and the exposed seeds are more often noticed.

This native semitropical weed grows along roadsides, and in hammocks and waste areas of peninsular Florida. It occurs along the Gulf area to Texas and is probably also in west Florida.

V-5A/SOL/SRY
(L 176-5-1)
CUCURBITACEAE

BJT

179. Begonia

Begonia cucullata Willdenow

These succulent, tropical, monoecious ornamentals, often grown as house plants further north, have alternate, glossy, strongly oblique leaves 5–7 cm wide and variable but showy staminate and pistillate flowers.

These annual or perennial herbs, native of South America, are naturalized on low roadsides and waste places of south Florida.

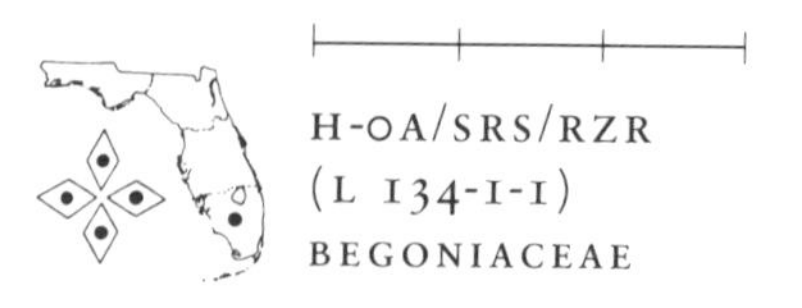

H-OA/SRS/RZR
(L 134-1-1)
BEGONIACEAE

EAH

180. False-teeth

Capparis flexuosa (L.) Linnaeus

The simple, alternate, pale green leaves of this shrub or small tree are 4–16 cm long and may be linear to widely ovate. The few flowers in the terminal inflorescence are white or pink but the slender fleshy fruits, 10–15 cm long, with conspicuous, exposed white seeds when ripe give the plant its colorful and appropriate common name.

The tropical Bay-leaved Caper Tree is infrequent in coastal hammocks and marl areas of south Florida and may occur as far north as Brevard County.

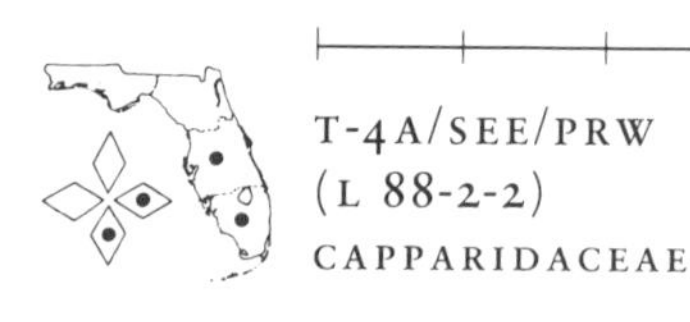

T-4A/SEE/PRW
(L 88-2-2)
CAPPARIDACEAE

BJT

181. Sea Rocket

Cakile lanceolata (Willd.) Schulz

These stout, branched succulent annual or biennial herbs are quite variable in leaf shape and size, but the small flowers, 1 cm wide, and the elongate, angled fruits aid in its identification. (The equally variable and more widespread *C. edentula*, which may have light lavender flowers, occurs in central and north Florida.)

Sea Rocket is a frequent and characteristic plant of beach dunes of the Atlantic coast from the Keys to the Carolinas.

H-4A/SEL/RRW
(L 87-7-1)
BRASSICACEAE

WSJ

182. Pepper Grass

Lepidium virginicum Linnaeus

The flowers of this annual herb are minute, and thus Pepper Grass is identified by the flat, round, pungent fruits which are 2–3 mm long. The fruits can be used to season salads and may have some value as food for wildlife.

A common weed of roadsides, old fields, and waste areas throughout Florida and much of the United States.

H-4A/SED/RRW
(L 87-8-1; R 88-4-3)
BRASSICACEAE

WSJ

183. Watercress

Nasturtium officinale R. Brown

The fleshy stems of this edible, introduced aquatic may be a meter or more long and produce roots from each node, or point of leaf attachment, when the plant is on mud banks or moist sand. The pinnately cut leaves have three to nine rounded segments. The slender pods are 1–2 cm long.

Cultivated and widely established in streams, springs, and pools throughout Florida and much of the United States.

A-4A/PBC/RRW
(L 87-6-1; R 88-20-1)
BRASSICACEAE

WSJ

184. Wild Radish

Raphanus raphanistrum Linnaeus

The flowers of Wild Radish, 2–3 cm across, are usually pale yellow, but they may also range from white to pink or lavender.

This often rank, introduced, annual weed is common in disturbed areas, fallow fields, and along roadsides from the Lake Okeechobee area north through northern Florida and on into Virginia.

H-4A/PBC/RRY
(R 88-10-1)
BRASSICACEAE

BJT

185. Waltheria

Waltheria indica Linnaeus

The ovate, stellate pubescent, coarsely serrate leaves of these shrubs or coarse herbs are 2–7 cm long. The compact axillary inflorescences and yellow flowers are characteristic. (The related Chocolate Weed, *Melochia corchorifolia*, appears to be similar but has lavender flowers; the unrelated but vegetatively similar *Malvastrum corchorifolium*, in the Malvaceae, has larger flowers.)

An occasional tropical component of the flora found in open pinelands, primarily of south Florida and the Keys, and rarely in central Florida.

H-5A/SOS/KRY
(L 122-3-1)
STERCULIACEAE

BJT

186. Indian Mallow

Abutilon permolle (Willd.) Sweet

The stems and leaves of this coarse herb or low shrub are densely stellate pubescent. The axillary flowers are 2–3 cm across.

A tropical species also found in disturbed areas along the south Florida coast and the Keys.

S-5A/SCS/SRY
(L 121-12-1)
MALVACEAE

DNP

187. Poppy Mallow

Callirhoe papaver (Cav.) Gray

The erose, or slightly toothed petals, the large poppylike flowers 5–8 cm in diameter, and the three to five parted leaves, with linear to elliptic segments, help separate these slender herbs from our other mallows.

Poppy Mallow, rare in our area, is known from scattered localities in northern Florida; it is native to sandy woods along the Gulf to Texas.

H-5A/SOL/SRR
MALVACEAE

CRB

188. Swamp Mallow

Hibiscus moscheutos Linnaeus

The conspicuous flowers of these shrubby perennial herbs are 10–15 cm across; the numerous anthers are united into a column around the style, a characteristic of the Mallow family. The lanceolate, serrate, tomentose leaves are 10–15 cm or more long.

A native of the southeastern United States found in open low grounds and brackish marshes of northern Florida.

S-5A/SES/SFW
(R 122-8-2)
MALVACEAE

BJT

EAH

189. Salt Marsh Mallow

Kosteletzkya virginica (L.) Presl ex Gray

A hairy herbaceous perennial (or small shrub in frost-free areas) 1–2 meters tall. Each flower is 5–8 cm across and lasts for only one day. (The closely related *K. pentasperma*, found only in south Florida, has white petals and bristly stinging hairs on the leaves and stems.)

Native in salt marshes and low sandy areas near the coast from Florida to Louisiana and New York.

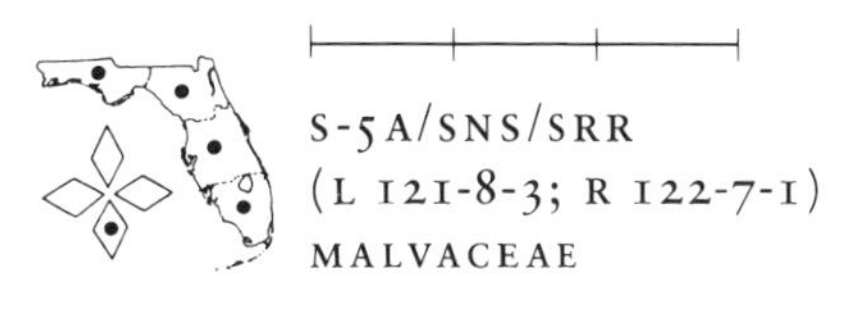

S-5A/SNS/SRR
(L 121-8-3; R 122-7-1)
MALVACEAE

190. Turk's-cap Mallow

Malvaviscus arboreus Cavanilles

Also called Wax Mallow, the pendant flowers of these bushy shrubs never open fully: their petals overlap to form a loose tube, about 3 cm long, with the stamen column protruding.

A tropical landscape plant persisting around old home sites, along roadsides and in waste places of peninsular Florida and Escambia County at the end of the Panhandle.

S-5A/SOD/STR
(L 121-7-1)
MALVACEAE

BJT

191. Caesar Weed

Urena lobata Linnaeus

These rank herbs or shrubs, to 3 meters tall, have widely ovate, shallowly lobed leaves 5–10 cm long that are densely stellate pubescent beneath. The five-parted flower and the dry fruit, which splits into five parts, are typical of the Mallow family.

A common tropical weed of disturbed soils and waste areas of south and central Florida.

S-5A/SOL/SRR
(L 121-4-1)
MALVACEAE

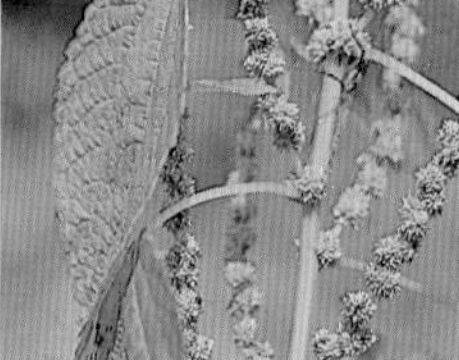

WSJ

192. False Nettle

Boehmeria cylindrica (L.) Swartz

As the name implies, these perennial herbs, with opposite, widely lanceolate, serrate leaves up to 1 dm long, look much like the related Wood Nettle (*Laportea canadensis*, rare in west Florida), which has stinging hairs. The compact clusters of minute flowers form a cylinder on small, essentially leafless, axillary branches.

A rather frequent weed of moist low ground and boggy areas throughout Florida and the Southeast.

H-OO/SNS/KAG
(L 65-2-1; R 59-3-1)
URTICACEAE

CRB

193. New Jersey Tea

Ceanothus americanus Linnaeus

The alternate, elliptic to lanceolate leaves of this low bushy shrub are 5–8 cm long and are pubescent beneath. The terminal flower clusters or inflorescences are longer than the leaves, which were once dried and used for tea.

A relatively frequent shrub of open deciduous woods and dry clearings from the Lake Okeechobee area north and west over much of eastern North America.

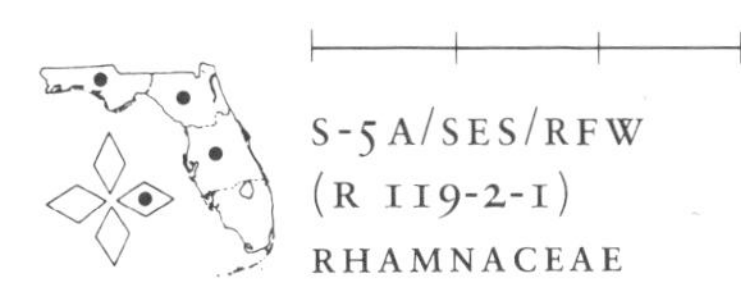

S-5A/SES/RFW
(R 119-2-1)
RHAMNACEAE

BJT

194. Spurge

Chamaesyce hypercifolia (L.) Millspaugh

The finely serrate, elliptic leaves of this erect spurge are 1–3.5 cm long. The flowers, greatly reduced and without petals, are borne together in a small inflorescence called a cyathium. The white "petals" are specialized glands. (The similar and related *C. hyssopifolia*, another of our sixteen species of Spurge, is a weed throughout the southeastern United States.)

A frequent plant of pinelands and disturbed areas of south and central Florida.

H-40/SES/KRW
(L 108-15-1)
EUPHORBIACEAE

RCS

195. Tread Softly; Stinging Nettle

Cnidoscolus stimulosus (Michx.) Engelmann & Gray

These low, rhizomatous, monoecious perennials are armed with numerous stinging hairs. The fragrant flowers, about 1 cm broad, have no petals, but the sepals are petaloid and showy. The palmately lobed leaves are 10–20 cm wide.

Frequent on sandhills, beach dunes, and in open pinelands in all sections of Florida and along the coast to Texas and Virginia.

H-5A/SOL/RFW
(L 108-17-1; R 107-1-1)
EUPHORBIACEAE

BJT

196. Pineland Croton

Croton linearis Jacquin

The narrow, 4–7 cm long, linear leaves of this small dioecious shrub have a yellow brown stellate pubescence beneath but are smooth on the upper surface. The racemes of staminate flowers are 4–10 cm long, the pistillate racemes mostly 4–5 cm long. (The wider leaves of the more common *C. glandulosus* and the rare *C. lobatus*, both found in our area, are not entire.)

Pineland Croton is found in the open coastal pinelands of south Florida and also in the West Indies.

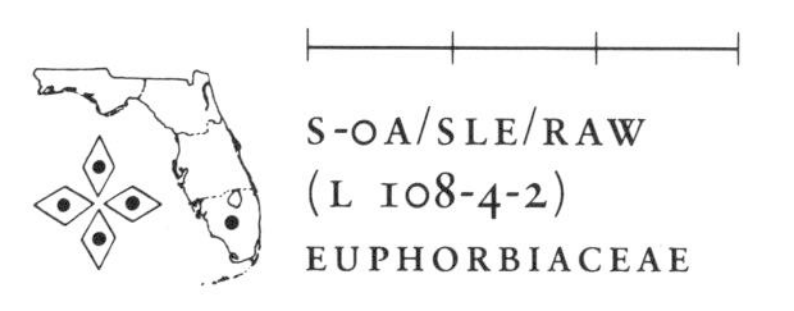

S-OA/SLE/RAW
(L 108-4-2)
EUPHORBIACEAE

BJT

197. Painted Leaf

Poinsettia heterophylla (L.) Klotzsch & Garcke

The wide upper leaves, either entire or coarsely lobed, are just below the inflorescence of the small inconspicuous flower clusters, or cyathia and are often partly or entirely red in this wild relative of our colorful cultivated Poinsettias. (The related *P. pinetorum*, endemic to south Florida, has narrow, linear leaves.)

A widespread perennial of moist clearings and pinelands throughout Florida and much of the United States.

H-OA/SEL/KAY
(L 108-11-1; R 107-11-2)
EUPHORBIACEAE

CRB

198. Poinsettia

Poinsettia pulcherrima (Willd. ex Klotzsch) Graham

This familiar tropical horticultural plant, only grown in greenhouses farther north, is widely planted as an ornamental in south Florida, where older shrubs may reach a height of 2 meters or more. The brilliant red leaves are below the small yellow flowers.

Plants of Poinsettia may persist around old homesites or garden areas in south Florida and near the coast in central Florida but are killed by freezing temperatures.

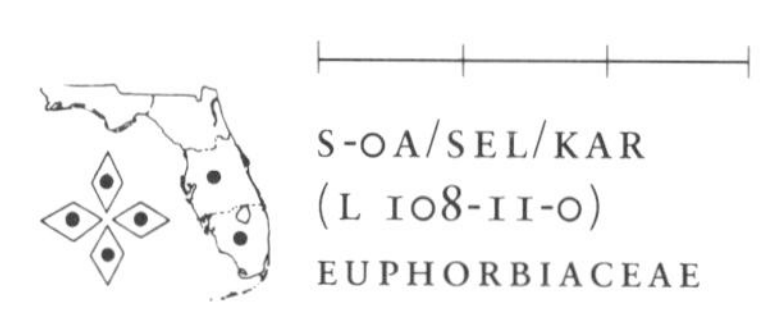

S-OA/SEL/KAR
(L 108-11-0)
EUPHORBIACEAE

BJT

199. Castor Bean

Ricinus communis Linnaeus

These rank, arborescent, monoecious perennial or annual tropical herbs may be 3–4 meters tall. The reddish, palmately lobed leaves are up to 4 dm long, and the large terminal inflorescences produce the conspicuous three-lobed fruits, 1.5–2 cm long, that bear the smooth, mottled, poisonous seeds from which castor oil is obtained.

In our area this widely cultivated introduced ornamental often escapes, or persists, around gardens, building sites, and old fields until killed by frost.

S-OA/SOL/RAG
(L 108-3-1; R 107-5-1)
EUPHORBIACEAE

WSJ

200. Mountain Spurge

Pachysandra procumbens Michaux

This botanically interesting, low evergreen perennial herb bears the thick ovate to obovate leaves and the 2–10 cm long spikes of brownish, musk-scented, monoecious flowers on separate branches. The flowers at the top of the spike are male, or staminate, and those at the bottom are female, or pistillate.

A widespread but rare plant of rich calcareous woods of Jackson County in west Florida; also found into Louisiana and north to West Virginia.

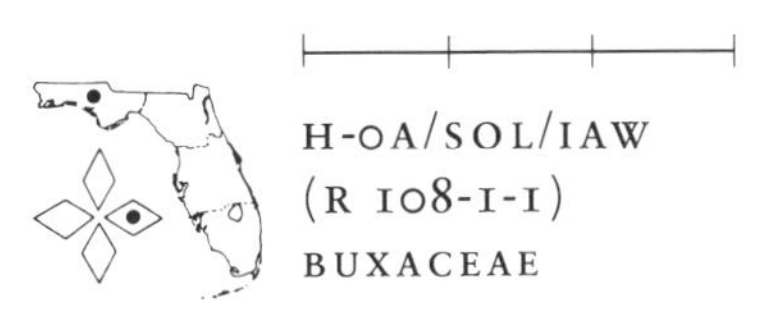

H-0A/SOL/IAW
(R 108-1-1)
BUXACEAE

RCS

201. Angel Trumpet

Datura innoxia P. Miller

These annuals are often cultivated for their large and very fragrant flowers. The branched stems grow to 2–3 meters or more in height, and the tubular flowers are 15–20 cm long. All parts of the plant are considered poisonous. (The African *D. metel*, with violet flowers, may also occasionally escape in our area.)

Angel Trumpets are occasionally escaped and naturalized along roadsides and in waste places in central and northern Florida.

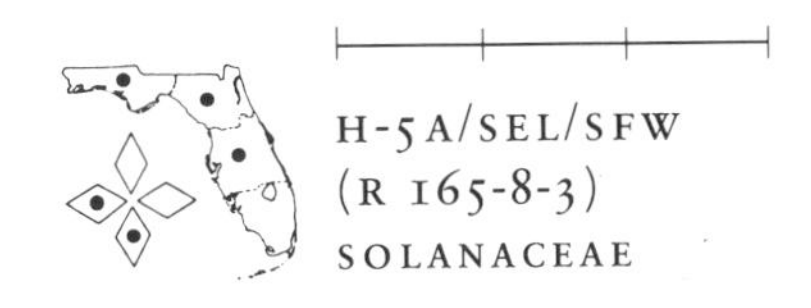

H-5A/SEL/SFW
(R 165-8-3)
SOLANACEAE

WSJ

202. Jimson Weed

Datura stramonium Linnaeus

The flowers of this rank, odorous native annual are 5–8 cm long and may be lavender to white. The thorny capsules, 3–5 cm or more in diameter, are often used in dried arrangements; however, both the seeds and leaves are very poisonous if ingested.

A locally frequent weed of barn lots and waste places from central Florida north over most of the eastern United States.

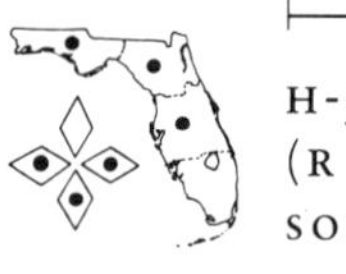

H-5A/SEL/SFB
(R 165-8-1)
SOLANACEAE

BJT

203. Matrimony Vine

Lycium carolinianum Walter

The spiny branches of this woody shrub are erect or spreading and may be up to 2 meters long. The rotate flowers are about 1 cm in diameter and the fleshy red berry about 1 cm in diameter.

Locally frequent along coastal dunes, marshes, and shell mounds of Florida and adjacent states.

S-5A/SLE/SFB
(L 167-1-1; R 165-1-1)
SOLANACEAE

BJT

204. Narrow Leaf Ground Cherry

Physalis angustifolia Nuttall

The extremely narrow leaves, five to ten or even twenty times greater in length than in width, immediately identify this perennial Ground Cherry. The flower is about 2 cm across, and the expanded, papery, ovoid calyx that encloses the round berry is about 2 cm long.

A plant of open coastal soils from south Florida to Louisiana.

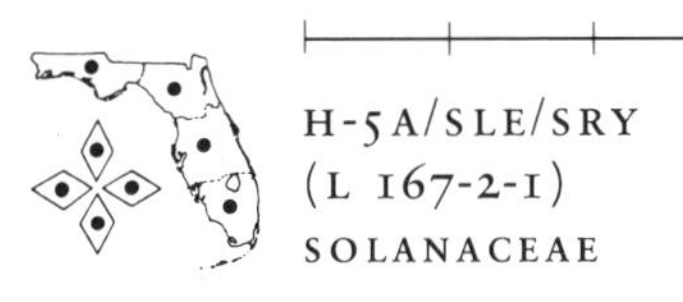

H-5A/SLE/SRY
(L 167-2-1)
SOLANACEAE

WSJ

205. Seaside Ground Cherry

Physalis viscosa Linnaeus

The stems and ovate to elliptic leaves of these trailing, rhizomatous perennials are stellate pubescent. The rotate-campanulate flowers are about 2 cm across and the papery, expanded fruiting calyx is 2.5–3.5 cm long.

Native to coastal dunes and roadsides from south Florida to Texas and the Carolinas.

H-5A/SEE/SRY
(L 167-2-2; R 165-3-1)
SOLANACEAE

WSJ

206. Common Nightshade

Solanum americanum P. Miller

An unarmed glabrous annual that branches freely to form a plant up to 1 meter tall and broad. The small white flowers are borne in clusters of three to five. The fleshy black berries, about 5 mm in diameter, are poisonous when green but edible when fully ripe. (The very similar *S. gracile* may also occur on coastal dunes and along brackish marshes of our area.)

This variable weedy native is frequent in waste places and disturbed areas from southern Florida north through much of the eastern United States.

H-5A/SOE/UFW
(L 167-3-8; R 165-5-9)
SOLANACEAE

BJT

207. Horse Nettle

Solanum carolinense Linnaeus

This is a widespread and weedy rhizomatous native perennial with coarse, prickled stems 0.5–1 meter tall. The white or lavender flowers are 2–3 cm across and the round yellow berries 1–2 cm in diameter. (The western *S. rostratum*, with longer spines and bright yellow flowers, is also becoming established in our area.)

Frequent in disturbed areas and waste places from central Florida north through much of the eastern United States.

H-5A/SOL/RRB
(R 165-5-5)
SOLANACEAE

BJT

208. Nightshade

Solanum diphyllum Linnaeus

The smooth, unarmed stems and the glabrous, elliptic leaves 6–8 cm long help identify this introduced shrub, which may be a meter tall and which bears round fruits that become a rich yellow when ripe.

Infrequently naturalized in hammocks and disturbed areas of south and central Florida.

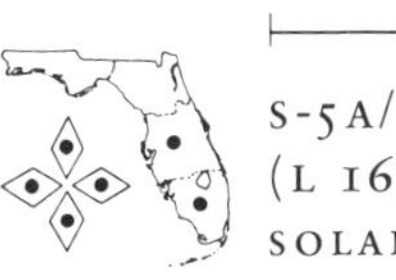

S-5A/SEE/PRB
(L 167-3-4)
SOLANACEAE

BJT

209. Blodgett's Nightshade

Solanum donianum Walpers

The stellate-hirsute pubescence of the branches and the 5–15 cm long elliptic leaves of this *Solanum* give it a grayish tone. (Indeed, the related but herbaceous *S. eleagnifolium*, introduced into Florida from the west, has similar pubescence and is called White Horse Nettle.)

A rare shrub of pinelands and hammocks of south Florida.

S-5A/SEE/PRW
(L 167-3-3)
SOLANACEAE

WSJ

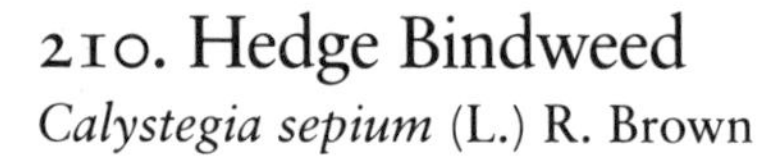

210. Hedge Bindweed

Calystegia sepium (L.) R. Brown

The funnelform flowers of *Calystegia*, 5–7 cm across, can be separated from those of *Convolvulus* by the two large leaflike bracts that conceal the calyx. The flowers of Hedge Bindweed may be white to pale rose purple. (The similar, but more northern *C. spithamaea* occurs in our state only in Jackson County.)

These perennial twining vines grow in open disturbed areas and along roadsides from central Florida north and west over much of the eastern United States.

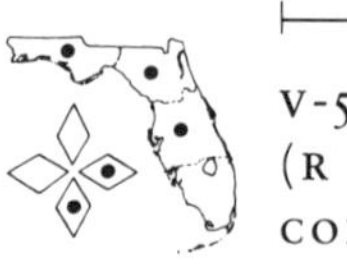

V-5A/SCE/SFW
(R 158-6-2)
CONVOLVULACEAE

DNP

211. Dodder; Love Vine

Cuscuta gronovii Willdenow

The flowers of these colorful, leafless twining vines are very small, usually only 1–3 mm long, and the several species are often difficult to separate. This species is most often found on woody hosts. (The similar *C. compacta* also grows on woody hosts while *C. campestris* is frequently on legumes or other herbs.)

Various species of these parasitic vines often occur in dense mats on host plants at scattered localities throughout Florida and much of the United States.

E-5N/---/KRW
(L 160-1-0; R 158-1-8)
CONVOLVULACEAE

BJT

212. Creeping Morning Glory

Evolvulus sericeus Swartz

The slender spreading or decumbent stems of these plants are only 1–3 dm long, and the white to pale blue corolla is about 1 cm in diameter. The stems and linear leaves are pubescent.

These small tropical perennials are infrequent in moist, open pinelands of south, central, and possibly north Florida into coastal Georgia and west to Mexico.

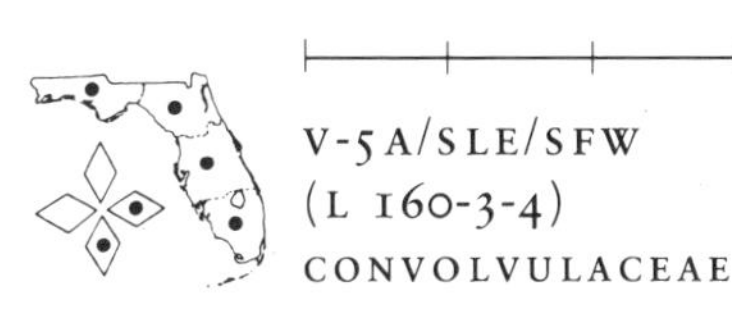

V-5A/SLE/SFW
(L 160-3-4)
CONVOLVULACEAE

BJT

213. Moon Flower

Ipomoea alba Linnaeus

The large flower of this twining, high-climbing vine is 10–15 cm long, and the ovate, entire to lobed leaves may be equally as long. The sepals are prolonged into hooked appendages, and the seeds of this species are essentially glabrous. (The closely related *I. tuba* has blunt sepals and the seeds are pubescent.)

This pantropical perennial occurs in coastal hammocks, and especially burned areas, of the Keys and north into central Florida.

V-5A/SOL/SFW
(L 160-10-1)
CONVOLVULACEAE

RCS

214. Scarlet Morning Glory

Ipomoea coccinea Linnaeus

The essentially unlobed crimson flowers of this tropical vine, which has ovate, angular, entire leaves, are 2–4 cm long and similar in size to those of Cypress Vine. (The ovate leaves of Cypress Vine, *I. quamoclit*, are pinnately dissected.)

Widely naturalized in Florida and much of the United States, these weedy but colorful vines may be found along railroads, fencerows, and in gardens and waste areas.

V-5A/SOE/UFR
(L 160-10-4; R 158-7-2)
CONVOLVULACEAE

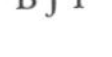

BJT

215. Morning Glory

Ipomoea indica (Burm. f.) Merrill

The leaves of this annual twining tropical vine are 5–10 cm long and may be entire or three-lobed. The white to pink flowers, 5–8 cm in diameter, may be solitary or two to three in a cluster. (The similar, widespread, weedy *I. purpurea* occurs in northern Florida.)

Native to hammocks and disturbed sites of peninsular Florida and the Keys.

V-5A/SCE/SFR
(L 160-10-10)
CONVOLVULACEAE

BJT

216. Glades Morning Glory

Ipomoea sagittata Poiret

The narrow sagittate leaves of this native twining vine are 3–10 cm long, and have distinctive linear to lanceolate, spreading, basal lobes.

These vigorous perennials often form dense mats on coastal dunes or over other vegetation along the margins of brackish marshes from south Florida along the coast to Texas and North Carolina.

V-5A/SCE/SFW
(L 160-10-5; R 158-7-9)
CONVOLVULACEAE

DNP

217. Beach Morning Glory

Ipomoea stolonifera (Cyr.) J. F. Gmelin

The leathery, or coriaceous, leaves of this glabrous, trailing perennial vine are oblong or sometimes three-lobed, 2.5–4.5 cm long, and the blunt apex is often notched. The flowers are 3.5–5 cm long or broad. (The succulent leaves of *I. pes-caprae*, which has lavender flowers, are ovate to orbicular or reniform.)

A frequent to occasional vine of coastal dunes of Florida and of other southeastern states.

V-5A/SOL/SFW
(L 160-10-9; R 158-7-5)
CONVOLVULACEAE

BJT

218. Jacquemontia

Jacquemontia curtissii Peter ex Hallier f.

This semiwoody, erect to prostrate vine, up to a meter long, has small variable leaves 1–2 cm long and a strongly five-lobed corolla 2–3 cm in diameter. (The very similar and closely related *J. reclinata* also occurs in south Florida, and *J. tamnifolia*, with pale blue flowers, occurs in central and northern Florida.)

This rare or overlooked perennial is endemic to the pinelands of south Florida.

V-5A/SOE/SFW
(L 160-12-3)
CONVOLVULACEAE

DNP

219. Cutleaf Morning Glory

Merremia dissecta (Jacq.) Hallier f.

The large ovate to lanceolate sepals, which become coriaceous and 2–4 cm long in fruit, the distinctive dissected lobes of the leaf, and glabrous seeds are characteristic of this twining vine.

Native but relatively infrequent in the pinelands of Florida; also found in Georgia and west to Texas.

V-5A/SOL/SFW
(L 160-8-1)
CONVOLVULACEAE

DNP

220. Trailing Morning Glory

Stylisma patens (Desr.) Myint

The trailing habit, the weakly lobed corolla 1 cm across and 1.5–2.5 cm long, and the elliptic to lanceolate leaves 2.5–7 cm long help separate these small flowered plants from those of the related *Jacquemontia*. (*S. villosa*, endemic to peninsular Florida, has villous, brown or tan pubescence.)

A frequent, small trailing vine of dry pinelands and scrub of central and northern Florida, and on the coastal plain into North Carolina and Mississippi.

V-5A/SLE/SFW
(R 158-3-4)
CONVOLVULACEAE

RCS

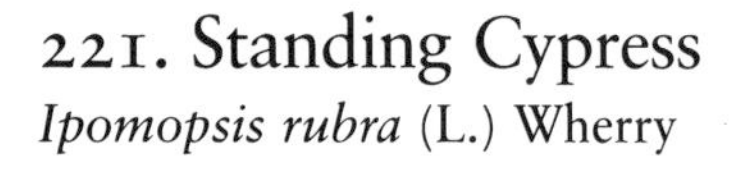

221. Standing Cypress

Ipomopsis rubra (L.) Wherry

The ovate leaves of these spectacular herbs, which grow to 1 meter tall, are pinnately divided into narrow linear segments. The tubular flowers, pollinated by hummingbirds, are about 2.5 cm long.

A sporadic biennial of dunes, sandhills, and waste places of central and northern Florida, and into Texas and the Carolinas.

H-5A/SOL/RTR
(R 159-3-1)
POLEMONIACEAE

CRB

222. Annual Phlox

Phlox drummondii Hooker

The alternate upper leaves, the annual habit, and the distinct "eye" in the center of each of the flowers, which may be white to deep pink and are 1–2 cm in diameter, are all distinctive features of this weedy but showy Phlox.

These glandular-pubescent horticultural plants, native of Texas, are naturalized in open, sandy waste places and along roadsides throughout Florida and coastal areas of the Southeast.

H-5O/SOE/URR
(L 161-1-1; R 159-1-1)
POLEMONIACEAE

CRB

223. Creeping Phlox

Phlox nivalis Loddiges ex Sweet

The slender, trailing, woody stems of these low perennials form compact tufts. The needlelike leaves of this Phlox are only 1–2.5 cm long, and the usually pink flowers are 1.5–2 cm across.

Creeping Phlox, found in open fields and pinelands, has a sporadic distribution from the general vicinity of Orlando north into Alabama and the Carolinas.

H-5O/SLE/URR
(R 159-1-2)
POLEMONIACEAE

CRB

224. Cardinal Flower

Lobelia cardinalis Linnaeus

Few native plants have flowers with such intense color. The flowers are 3–4 cm long, strongly two-lipped (bilaterally symmetrical, or zygomorphic), and are pollinated by hummingbirds.

These spectacular herbaceous perennials may reach a height of 1.5 meters in moist open meadows and along stream banks in central and northern Florida and throughout the eastern United States.

H-5A/SES/RZR
(R 178-6-1)
CAMPANULACEAE

RCS

225. Glades Lobelia

Lobelia glandulosa Walter

The strongly three-lobed lower lip of these attractive, strongly zygomorphic flowers is 1 cm wide; the throat of the corolla is pubescent. (We have several other similar blue Lobelias; however, *L. puberula* has puberulent stems and *L. paludosa* has much reduced upper leaves.)

Despite the common name given these plants in south Florida, they are found in wet pinelands and swamps in all sections of Florida and northward to Virginia.

H-5A/SNS/RZB
(L 177-2-3; R 178-6-5)
CAMPANULACEAE

BJT

226. Venus' Looking-glass

Triodanis perfoliata (L.) Nieuwland

The flowering stems of this native annual are 1.5–4.5 dm tall and are essentially encircled by the base of each sessile, heart-shaped leaf. The flowers are solitary in the axils of the upper leaves; the conspicuous, open purple flowers are cross-pollinated by insects; the inconspicuous cleistogamous, or closed, flowers never open and self-pollinate.

A rather attractive common weed of fallow fields, gardens, roadsides, and waste areas of central and northern Florida and other southeastern states.

H-5A/SOS/SRB
(R 178-1-1)
CAMPANULACEAE

WSJ

227. Wahlenbergia

Wahlenbergia marginata (Thunb.) de Candolle

The flowers of these slender, introduced perennials are only 1 cm or less across, but there are often many stems, 1–6 dm tall, on each plant, and thus large populations can be rather colorful.

A harmless Asiatic weed of dry sandy clearings, roadsides, and waste places from near Orlando north to the Carolinas and west to Alabama.

H-5A/SLE/PRB
(R 178-3-1)
CAMPANULACEAE

RCS

228. Mistletoe

Phoradendron serotinum (Raf.) M.C. Johnston

These small, parasitic, evergreen shrubs have brittle stems and thick, opposite, obovate leaves 2–6 cm long. These plants are dioecious, and thus the white berries, often used for holiday decoration, form only on the female, or pistillate, plants.

A common parasite that grows on the upper branches of many species of trees, but especially frequent on oaks, in a variety of habitats throughout Florida and much of the eastern United States.

E-OO/SBE/SAG
(L 67-1-2; R 61-1-1)
LORANTHACEAE

BJT

229. Fringe Tree

Chionanthus virginicus Linnaeus

Because of its showy panicles of white flowers and the attractive clusters of fleshy, dark blue fruits, several varieties of this native shrub have been developed for landscape use. The elliptic to ovate deciduous leaves may be up to 2 dm long; the four linear petals of each flower are 1.5–2.5 cm long.

A frequent shrub of rich woods, stream banks, and savannas of central Florida, north and west over much of the Southeast.

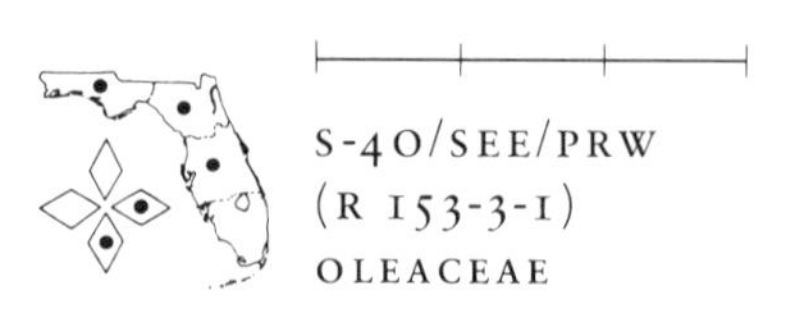

S-40/SEE/PRW
(R 153-3-1)
OLEACEAE

BJT

230. Puncture Weed

Tribulus cistoides Linnaeus

The showy flowers of this prostrate weed are 4–5 cm across and the pinnately compound leaves are 4–6 cm long. The hard fruits have a few strong spines, which give the plant its common name.

This European perennial is naturalized along sandy roadsides and in disturbed areas in all parts of Florida and into Georgia and other Gulf Coast states to Texas.

H-50/PEE/SRY
(L 100-2-1)
ZYGOPHYLLACEAE

BJT

231. Oxalis; Sorrel

Oxalis dillenii Jacquin

The three heart-shaped leaflets of each trifoliate leaf, the round or radially symmetrical flowers (about 1 cm across in this species), and the pungent flavor of oxalic acid when a leaf is chewed, are characteristic of Oxalis. (Two similar species in our area, *O. corniculata*, and *O. stricta*, also have yellow flowers; *O. violacea* has violet flowers.)

One or more of these low erect perennials are common in lawns, old fields, disturbed areas, and along roadsides throughout Florida and other states of the central and eastern United States.

H-5A/TBE/UFY
(L 97-1-1; R 100-1-5)
OXALIDACEAE

BJT

232. Crane's Bill

Geranium carolinianum Linnaeus

This low, branched, pubescent, weedy annual is the only representative of the Geranium family to be found in Florida. The flowers may be pink or white and are about 1 cm across; the palmately lobed leaves are 4–8 cm wide.

Frequent to common in old fields and disturbed areas more or less throughout Florida and the United States.

H-5A/SOL/SRR
(L 96-1-1; R 101-1-4)
GERANIACEAE

BJT

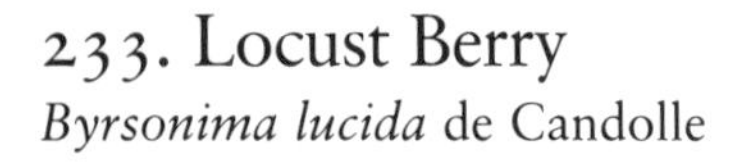

233. Locust Berry

Byrsonima lucida de Candolle

The opposite, obovate, evergreen leaves of these shrubs or small trees are 2–5 cm long; the white to pink petals may become yellowish with age giving the plant an attractive multicolored appearance when in bloom.

A plant of south Florida pinelands which also occurs in the West Indies.

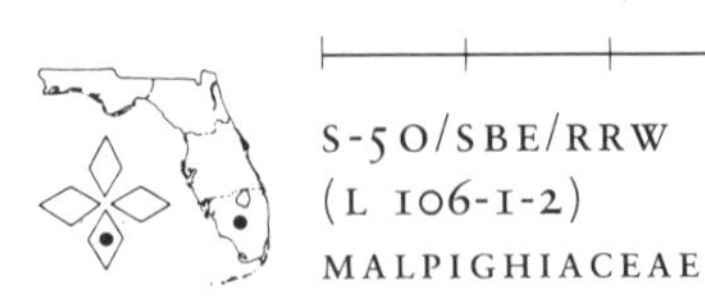

S-50/SBE/RRW
(L 106-1-2)
MALPIGHIACEAE

BJT

234. White Bachelor-button

Polygala balduinii Nuttall

The compact racemes of this branched herb may be 2–3 cm long. The smooth, obovate to lanceolate leaves are entire.

A conspicuous and relatively frequent plant of wet pinelands in all sections of Florida and on into Georgia and Mississippi.

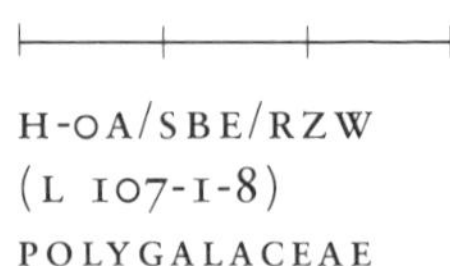

H-OA/SBE/RZW
(L 107-1-8)
POLYGALACEAE

WSJ

235. Drum-heads

Polygala cruciata Linnaeus

The whorled leaves, three or four at a node, and the compact, greenish rose raceme 1–4 cm long help identify this erect annual.

A frequent component of the flora of wet pinelands, bogs, and pocosins of Florida, much of the Southeast, and beyond.

H-OW/SLE/RZR
(L 107-1-1; R 106-1-10)
POLYGALACEAE

WSJ

236. Tall Milkwort

Polygala cymosa Walter

The basal rosette of linear to lanceolate leaves, 4–7 cm long, and the cymosely branched flowering stalk to 1 meter tall are characteristic of this Milkwort. (The basal leaves of the similar *P. ramosa* are elliptic to spatulate and are often withered when the plant blooms.)

This glabrous biennial is frequent in the moist pinelands of Florida and along the coast to Louisiana and Delaware.

H-OB/SNE/RZY
(L 107-1-3; R 106-1-17)
POLYGALACEAE

BJT

237. Large-Flower Polygala

Polygala grandiflora Walter

The two relatively large, pink petaloid sepals, 5–7 mm long and which form a pair of "wings" with the small corolla between them, are clearly seen on the large, uncrowded flowers of this species. The slender alternate leaves are 2–5 cm long.

This perennial Polygala occurs in moist pinelands throughout Florida, north to the Carolinas, and west to Louisiana.

H-OA/SNE/RZR
(L 107-1-12; R 106-1-3)
POLYGALACEAE

BJT

238. Bog Bachelor-Button; Candyweed

Polygala lutea Linnaeus

This colorful native biennial grows to 3 dm or more in height and has compact racemes, about 2 cm in diameter, that bloom throughout the summer; the small fruits fall as they ripen leaving scars on the flower stalk beneath the current blooms.

A coastal plain species that is fairly common along moist roadsides, on low pinelands, and in sandy bogs throughout the state and much of the Southeast.

H-OA/SEE/RZY
(L 107-1-6; R 106-1-14)
POLYGALACEAE

BJT

239. Bachelor-button

Polygala nana (Michx.) de Candolle

The compact racemes of this low, glabrous, rather succulent biennial are 2–3 cm long; the bright yellow flowers turn green or blue green on drying.

A southeastern herb that is relatively frequent in wet pinelands and moist, open sandy areas throughout Florida, west to Texas and north to the Carolinas.

H-OB/SBE/RZY
(L 107-1-7; R 106-1-15)
POLYGALACEAE

RCS

240. Polygala

Polygala polygama Walter

The large flowers and the more open racemes of this Milkwort are somewhat similar to those of *P. grandiflora*, but the narrower "wings" and the fringed petals help identify this species, which, interestingly, also produces small unopened, or cleistogamous, self-pollinating flowers underground.

Widespread over much of the eastern United States, these woodland perennials, or biennials, are found in dry pinelands and dune areas in all sections of Florida.

H-OA/SBE/RZR
(L 107-1-11; R 106-1-2)
POLYGALACEAE

BJT

241. Sweet Orange

Citrus sinensis (L.) Osbeck

The Orange blossom is the State Flower of Florida. The waxy, fragrant flowers of Sweet Orange, and of other *Citrus* species (all of which are separated on fruit characters), are about 2 cm wide and have a distinctive ring of stamens.

Seven species of *Citrus*, all introduced, are grown horticulturally or agriculturally in Florida. All have become naturalized to some extent, or persist from prior cultivation, and may be found in varying frequency in hammocks, disturbed areas, and old groves primarily in south and central Florida.

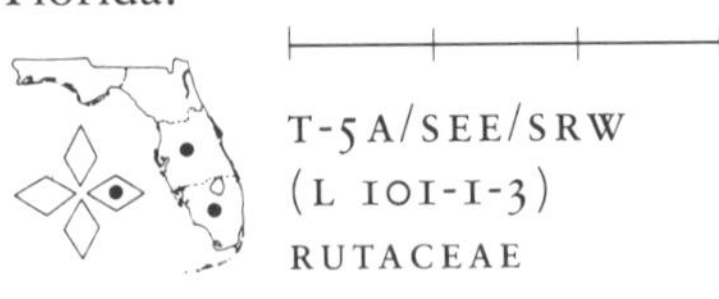

T-5A/SEE/SRW
(L 101-1-3)
RUTACEAE

WSJ

242. Hop Tree

Ptelea trifoliata Linnaeus

The trifoliate or three-parted leaf and the dry, round, flat, winged fruits, 1.5–2.5 cm broad, aid in the identification of these native shrubs or small trees that are related to the cultivated *Citrus* trees.

Sporadic in rich woods and along stream banks at scattered localities from Polk County north to New York and west to Texas.

T-4A/MEE/UAG
(R 103-2-1)
RUTACEAE

BJT

243. Bay Cedar

Suriana maritima Linnaeus

This densely branched shrub or small tree has thick, crowded, evergreen leaves only 1–4 cm long and rather inconspicuous flowers about 2.5 cm across.

Widespread along the dunes and in the maritime thickets of the new world tropics, this coastal shrub occurs at a reduced frequency in Florida to the approximate latitude of Tampa where it reaches its northern limit.

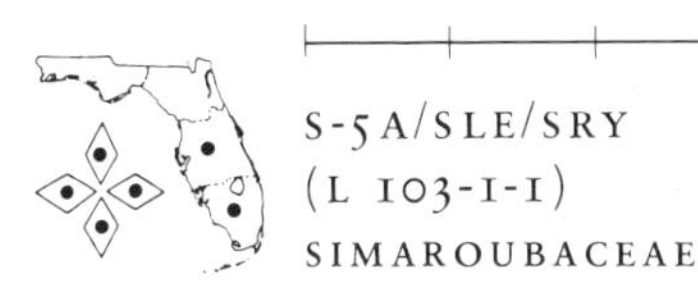

S-5A/SLE/SRY
(L 103-1-1)
SIMAROUBACEAE

HOW

244. Chinaberry

Melia azedarach Linnaeus

Chinaberry is easily identified by its bipinnately compound leaves and the large panicles of purple flowers or round yellow berries. The compact, rounded form of this small tree provides another common name: Umbrella Tree.

This frequent Asiatic horticultural plant is well naturalized in thickets, old fields, and disturbed areas throughout Florida and primarily in the coastal plain portions of other southeastern states.

T-5A/BES/PRB
(L 105-2-1; R 105-1-1)
MELIACEAE

BJT 

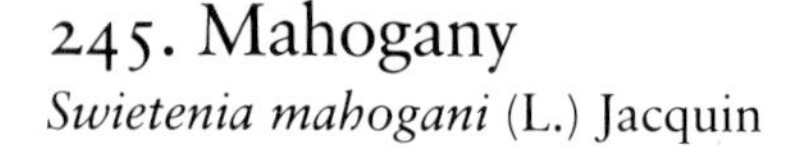

245. Mahogany

Swietenia mahogani (L.) Jacquin

These large trees, up to 25 meters tall, have pinnately compound leaves with four to eight glossy, elliptic leaflets 4–10 cm long. The large erect capsules, 6–12 cm long, open from the bottom to release the seeds.

A tropical tree that reaches its northern limit in the coastal hammocks of extreme south Florida and the Keys but is also widely planted throughout this area as an ornamental.

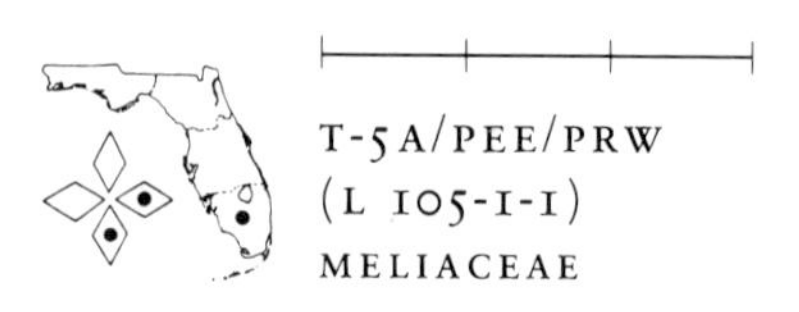

T-5A/PEE/PRW
(L 105-1-1)
MELIACEAE

BJT

246. Gumbo Limbo

Bursera simaruba (L.) Sargent

The characteristic, and attractive, part of this native deciduous tree is neither the flower nor the glossy green pinnate leaves but the smooth, lustrous bark of the older stems. Mature trees of Gumbo Limbo may be up to 25 meters tall and a meter in diameter; they produce a resin that is used medicinally.

A characteristic element of coastal hammocks and shell mounds of the Keys, and north to about Tampa, Gumbo Limbo is also planted as an ornamental.

T-0A/PNE/RAW
(L 104-1-1)
BURSERACEAE

BJT

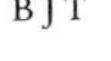

BJT

247. Poisonwood

Metopium toxiferum (L.) Krug & Urban

The alternate, pinnately compound leaves of this shrub or small tree have three to seven ovate, leathery, or coriaceous, leaflets. All parts of this native tree, and especially the resinous sap, are poisonous to touch.

Widespread in hammocks, pinelands, and on coastal dunes of south Florida and the Keys.

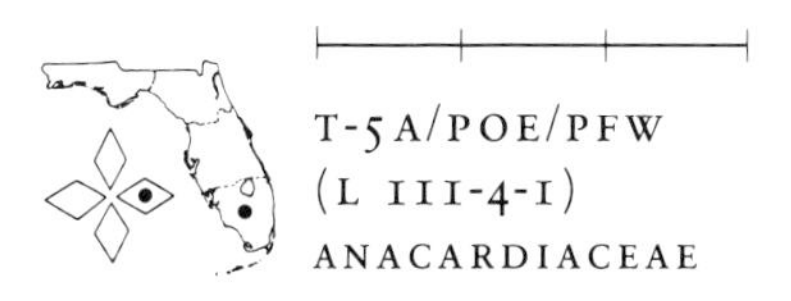

T-5A/POE/PFW
(L III-4-I)
ANACARDIACEAE

248. Winged Sumac

Rhus copallina Linnaeus

The rachis, or stalk, of the pinnately compound leaves of this nonpoisonous shrub or small tree has a narrow wing between the individual leaflets. The small red fruits are in compact terminal panicles 1–3 dm long. (The fruits of Poison Sumac, *R. vernix*, are white.)

A frequent native of dry roadsides, fence rows, thickets, and waste areas throughout Florida and the eastern United States.

S-5A/PNS/PAW
(L III-3-I; R IIO-I-7)
ANACARDIACEAE

EAH

249. Brazilian Pepper

Schinus terebinthifolius Raddi

Although this weedy, introduced shrub or small tree with its clusters of red hollylike berries is related to Poison Ivy and is toxic to some people, it is often used as Christmas decoration and, as indicated by the common name, a spice. However, until more is known about the toxic properties of this tropical plant, the latter use seems especially inappropriate and dangerous.

Brazilian Pepper, now common in waste areas, old fields, and hammocks of peninsular Florida, is invading the habitats of many of our native plants.

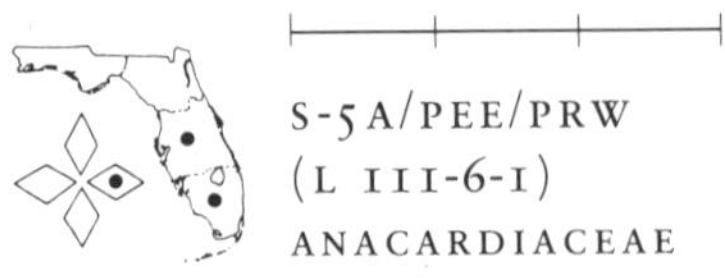

S-5A/PEE/PRW
(L III-6-1)
ANACARDIACEAE

BJT

250. Poison Ivy; Poison Oak

Toxicodendron radicans (L.) Kuntze

Contact with any part of these weedy, poisonous, woody vines results in severe skin irritation and blistering in many people. Fortunately, the three smooth, usually pointed leaflets, each 5–10 cm long, make the plant easy to recognize. The small clustered fruits are white.

A common perennial of open woods, pinelands, roadsides, and disturbed areas throughout Florida and the eastern United States.

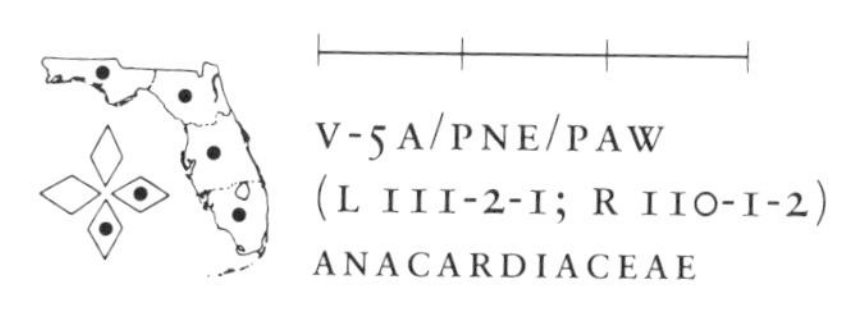

V-5A/PNE/PAW
(L 111-2-1; R 110-1-2)
ANACARDIACEAE

EAH

251. Balloon Vine

Cardiospermum halicacabum Linnaeus

The small, irregular white flowers of Balloon Vine are rather inconspicuous; it is the attractive, inflated, papery capsules, 2–4 cm long, that give the plant its common name.

This tropical, herbaceous, annual vine is widely cultivated and has become naturalized in a limited number of disturbed areas in peninsular Florida and several widely scattered localities in other southeastern states.

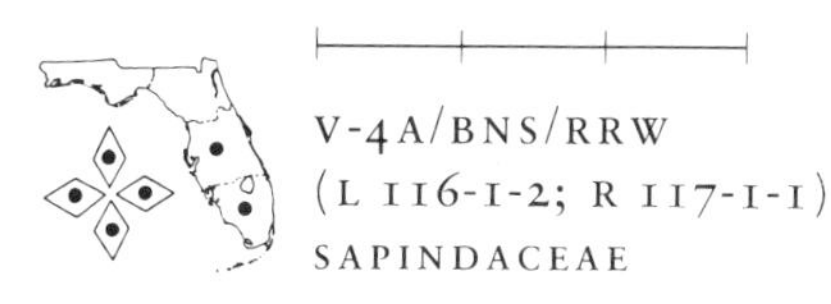

V-4A/BNS/RRW
(L 116-1-2; R 117-1-1)
SAPINDACEAE

EAH

252. Varnish Leaf

Dodonaea viscosa (L.) Jacquin

This dioecious, tropical shrub or small tree, which bears axillary clusters of small, pale yellow flowers, has viscid, generally oblanceolate evergreen leaves, 6–12 cm long. The strongly three-winged fruits are 1.5–2.5 cm across.

An infrequent inhabitant of coastal pinelands and hammocks from the Keys and south Florida north into central Florida to about Tampa.

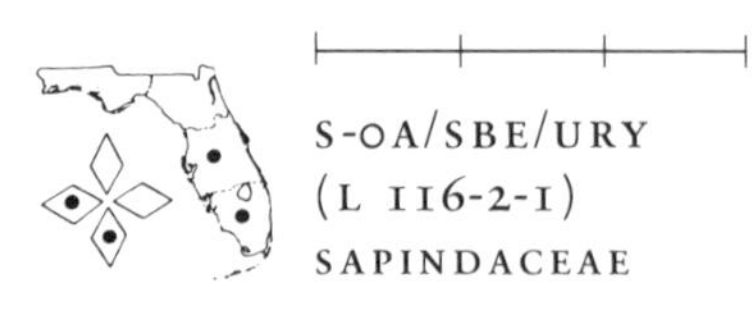

S-OA/SBE/URY
(L 116-2-1)
SAPINDACEAE

WSJ

253. Red Maple

Acer rubrum Linnaeus

The small but numerous rich red flowers and young fruits of Red Maple appear before the opposite, three-lobed leaves and add splashes of color to the low woodlands in late winter or very early spring.

This small to medium tree is common over much of the eastern United States and is frequent in low woods and along stream banks throughout Florida.

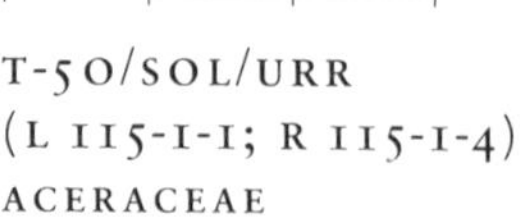

T-5O/SOL/URR
(L 115-1-1; R 115-1-4)
ACERACEAE

BJT

254. Red Buckeye

Aesculus pavia Linnaeus

The terminal raceme of 2–3 cm long red flowers and the opposite, palmately compound leaves make this native Buckeye easy to recognize. The large seeds are poisonous if eaten, but this colorful shrub or small tree is easily grown from fresh seed and flowers in 3 years.

Infrequent to common along woodland streams and swamp margins on the coastal plain from central Florida to Texas and the Carolinas.

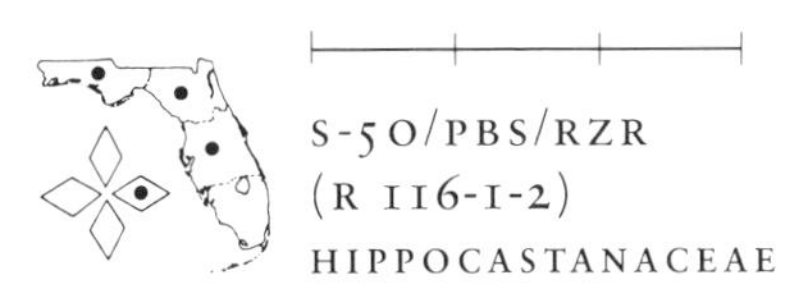

S-5O/PBS/RZR
(R 116-1-2)
HIPPOCASTANACEAE

CRB

255. Bayberry

Myrica cerifera Linnaeus

The aromatic, evergreen leaves of Wax Myrtle, as these bushy shrubs are also called, may be 4–10 cm long and are often toothed toward the apex. The wax-covered white berries, 2–3 mm in diameter, are the source of the fragrant Bayberry wax used in colonial times, and still today, by candle makers.

A common monoecious or dioecious shrub of hammocks, pinelands, marshes, and moist thickets in all sections of Florida and the coastal plain portions of other southeastern states.

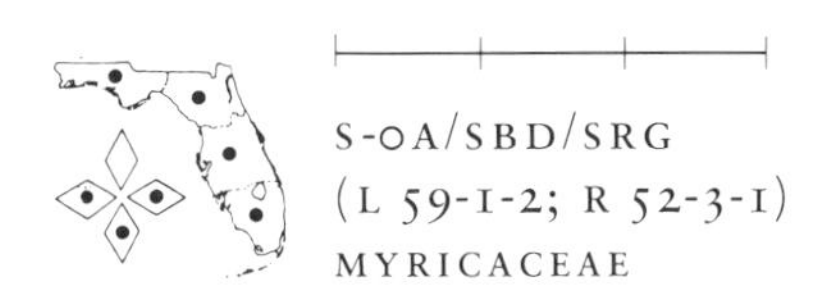

S-OA/SBD/SRG
(L 59-1-2; R 52-3-1)
MYRICACEAE

WSJ

256. Sea Purslane

Sesuvium portulacastrum (L.) Linnaeus

The petal-like sepals of these fleshy, prostrate or erect herbs are 1–2 cm long, and the pedicillate flowers have numerous stamens. (Flowers of the closely related *S. maritimum* are smaller, sessile, and have only five stamens.)

These native annuals are common on sea beaches and open dunes from Florida to New York and Mississippi.

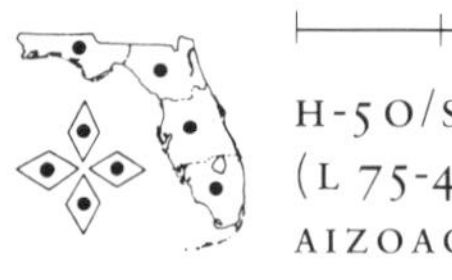

H-5O/SEE/SRR
(L 75-4-2; R 69-1-1)
AIZOACEAE

BJT

257. Prickly Pear

Opuntia humifusa (Raf.) Rafinesque

The fleshy, flattened segments of the prostrate green stems of this cactus are usually tightly jointed and not easily broken apart. The flowers, 5–7 cm in diameter, produce reddish, pulpy, obovoid fruits 2–3 cm long. (The stem segments of *O. triacantha* and *O. cubensis*, both restricted to the Keys, separate easily.)

A frequent perennial of sandy clearings, roadsides, and pinelands throughout the state, and on to Louisiana and the Carolinas.

H-9N/---/SRY
(L 135-1-3; R 132-1-1)
CACTACEAE

BCB

258. Pink Purslane

Portulaca pilosa Linnaeus

The slender, fleshy leaves of these prostrate, succulent annuals are usually opposite, or nearly so, and have a dense, axillary tuft of hairs, or trichomes. The corolla is about 1 cm across.

Pink Purslane is a frequent plant of sandy pinelands and waste places throughout Florida and on into Mississippi and the Carolinas.

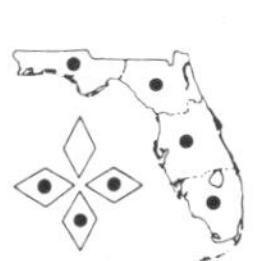

H-5O/SLE/SRR
(L 76-2-1; R 70-3-3)
PORTULACACEAE

WSJ

259. Mexican Tea

Chenopodium ambrosioides Linnaeus

This coarse aromatic perennial, with elliptic to lanceolate leaves 3–10 cm long, may be 1.5 meters tall and is beset with resinous glands or dots.

This cosmopolitan weed occurs throughout Florida in waste places and old fields.

H-OA/SND/PRG
(L 71-2-1; R 64-3-1)
CHENOPODIACEAE

BJT

260. Perennial Glasswort

Salicornia virginica Linnaeus

The prostrate stems of these rather stout, succulent perennials may be 5–6 dm long and have erect, green, leafless branches. The flower spikes, which become bright red in fruit, are 2–10 cm long. (The main stems of Annual Glasswort, *S. bigelovii*, also common in our area, are erect.)

Glassworts are a common element of the salt marsh flora along the entire east and west coastlines of the United States.

H-ON/---/IRG
(L 71-1-2; R 64-5-3)
CHENOPODIACEAE

BJT

261. Alligator Weed

Alternanthera philoxeroides (Mart.) Griesbach

The meter-long stems of this rank aquatic perennial, with opposite elliptic leaves 5–13 cm long, may be erect or trailing and readily root at the nodes, thus forming dense mats over small streams and lakes. (The related *A. maritima*, more tolerant of salt water, grows on sea beaches of south Florida.)

A common introduced weed of wetlands, streams, and ponds throughout Florida and the coastal plain to Louisiana and North Carolina.

A-5O/SEE/SRW
(L 72-4-3; R 66-1-2)
AMARANTHACEAE

BJT

262. Bloodleaf

Iresine diffusa Humboldt & Bonpland

The paniculate inflorescence of many minute flowers of this erect or sprawling, often vinelike, herb has a light, feathery appearance. The opposite leaves of *Iresine* are rather variable in general shape, size, and pubescence and may be 5–15 cm long. (The inflorescence of the related Cottonweed, *Froelichia floridana*, is much more compact.)

This annual, or weak perennial, is frequent to common in hammocks and disturbed areas throughout Florida and the coastal plain of other southeastern states.

H-00/SOS/PAW
(L 72-8-1)
AMARANTHACEAE

BJT

263. Chickweed

Stellaria media (L.) Villars

The five petals of the small flowers of these prostrate or decumbent annuals are only about 3 mm long and are so deeply cleft that they appear to be ten.

Chickweed is a common lawn and garden pest over much of the eastern United States and is relatively frequent in northern and central Florida down to about Orlando.

H-50/SOE/SRW
(R 71-7-2)
CARYOPHYLLACEAE

BCB

264. Coral Vine

Antigonon leptopus Hooker & Arnott

Coral Vine climbs by means of a tendril at the end of each colorful raceme or flower cluster. The rose sepals are about 2 cm long.

A native of Mexico that has escaped from cultivation and become naturalized along roadsides, hammocks, and waste places throughout peninsular Florida and has also been reported for northern Florida.

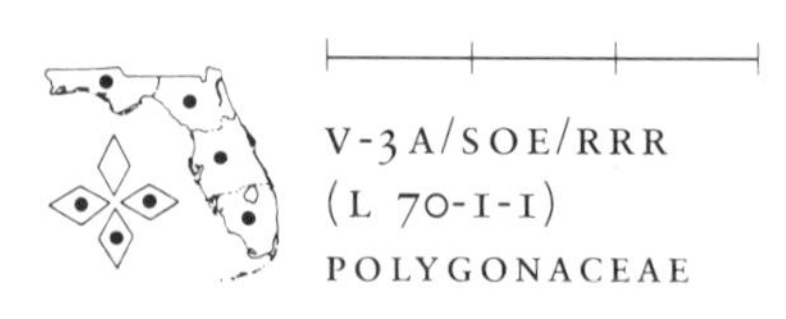

V-3A/SOE/RRR
(L 70-1-1)
POLYGONACEAE

BJT

265. Sea Grape

Coccoloba uvifera (L.) Linnaeus

A shrub or small tree with nearly round evergreen leaves 5–20 cm across. The reddish fruits are edible. (The leaves of the related Pigeon Plum, *C. diversifolia*, are ovate and longer than wide.)

The tropical Sea Grape is native to the coastal hammocks, dunes, and beaches of south Florida and the Keys.

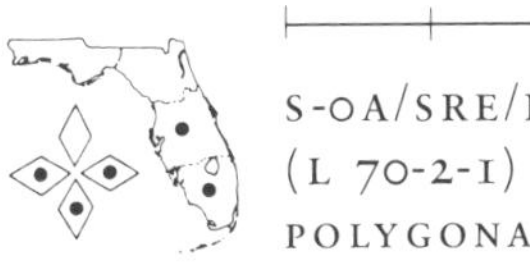

S-OA/SRE/RRW
(L 70-2-1)
POLYGONACEAE

DNP

266. Wild Buckwheat

Eriogonum tomentosum Michaux

These erect, pubescent perennials have a basal rosette of distinctive elliptic, long-petiolate leaves 7–12 cm long that are glabrous above but densely white or tan tomentose beneath. The spreading inflorescence is up to 1 dm across.

Dog-tongue, as this plant is also called, is frequent locally in dry pinelands and Turkey-oak sandhills of northern and central Florida, and into Alabama and South Carolina.

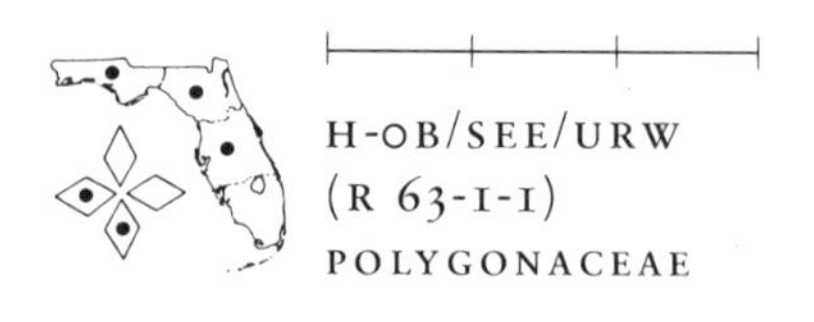

H-OB/SEE/URW
(R 63-1-1)
POLYGONACEAE

WSJ

267. Jointweed

Polygonella polygama (Vent.) Engelmann & Gray

The individual terminal racemes of Jointweed are 1–3 cm long. These erect, leafy-stemmed perennials may be 3–6 dm tall. (The related woody Wireweed, *P. myriophylla*, endemic to central Florida sand scrub, is prostrate; also, in *P. ciliata* and *P. gracilis* the stem leaves are early deciduous.)

A frequent plant of the Florida scrub and pinelands found on the coastal plain north to Virginia.

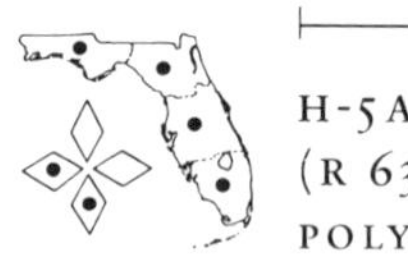

H-5A/SLE/RRW
(R 63-6-2)
POLYGONACEAE

GCP

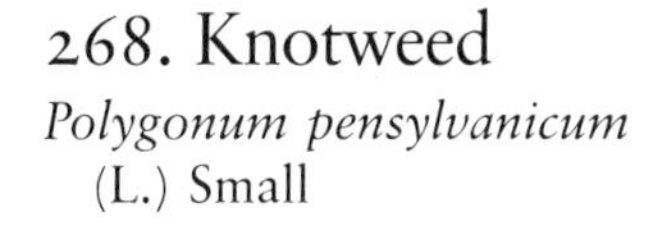

268. Knotweed

Polygonum pensylvanicum (L.) Small

The compact, knotted raceme, 3–10 cm long, and the alternate, lanceolate leaves with a membranous, expanded, sheath (ocrea) at the base of the petiole are characteristic of all members of this large genus. (Both *P. punctatum* with white, glandular-punctate sepals, and the perennial *P. setaceum* reach south Florida.)

One or more of our species of Knotweeds are common in swamps, marshes, and wet ditches in each section of Florida and over much of the eastern United States.

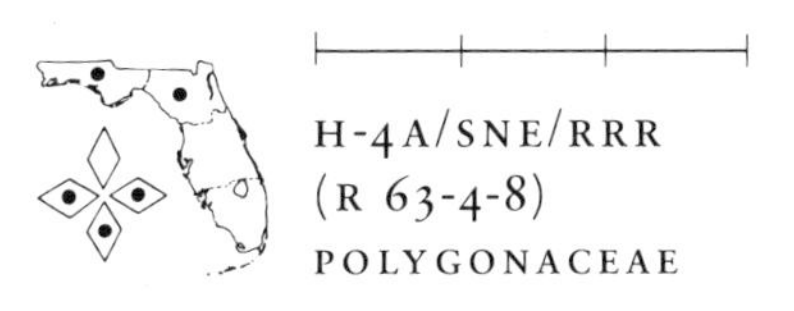

H-4A/SNE/RRR
(R 63-4-8)
POLYGONACEAE

RCS

269. Hastate Leaf Dock

Rumex hastatulus Baldwin ex Elliott

The narrow fruiting panicles or branching racemes of this dioecious weed are often 3–4 dm long on slender stalks 5–10 dm tall. (The leaves of the introduced Curled Dock, *R. crispus*, are not hastate at the base.)

This perennial often forms rather showy masses in old sandy fields of central and northern Florida and, on the coastal plain, into Mississippi and the Carolinas.

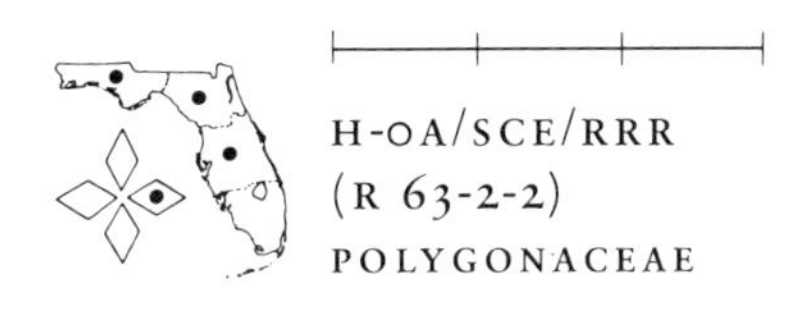

H-OA/SCE/RRR
(R 63-2-2)
POLYGONACEAE

BJT

270. Saltwort

Batis maritima Linnaeus

The opposite, sessile, fleshy, linear leaves of this strongly scented shrub are 1–3 cm long, and the minute flowers are in dense axillary spikes 1 cm long.

A common tropical perennial of salt marshes and other shore habitats, Saltwort is found in the United States from Florida to Mississippi and the Carolinas, often with *Salicornia*, or Glasswort.

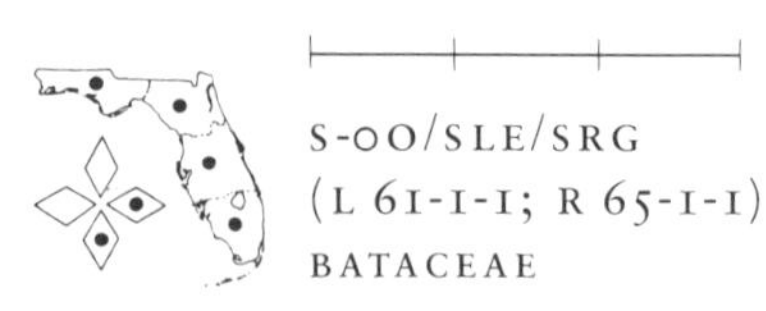

S-OO/SLE/SRG
(L 61-1-1; R 65-1-1)
BATACEAE

BJT

271. Australian Pine

Casuarina equisetifolia Linnaeus ex J.R. & G. Forster

The leaves of these large, handsome introduced trees are only small scales in a whorl around the dark green, slender, jointed, evergreen branches that give the tree, which has a small conelike fruit, the appearance of a pine.

Widely planted as a windbreak in peninsular Florida and now thoroughly naturalized, these Australian trees are spreading throughout the area.

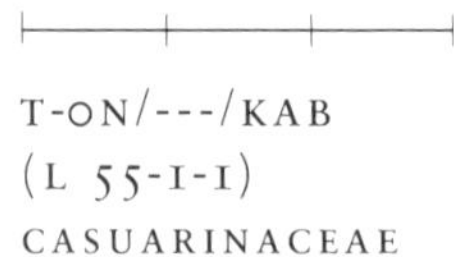

T-ON/---/KAB
(L 55-1-1)
CASUARINACEAE

CRB

CCF

272. Turkey Oak

Quercus laevis Walter

The obovate but deeply lobed leaves of these small, often stunted trees are 10–15 cm long. Although deciduous, the bright maroon to crimson leaves stay on the trees well into the fall or early winter.

These small trees are often the dominant vegetation, perhaps with Long-leaf Pine, on the dry Turkey Oak sandhills and ridges of Florida and other coastal states of the Southeast.

T-OA/SBL/IAY
(L 62-1-1; R 55-3-19)
FAGACEAE

WSJ

273. Live Oak

Quercus virginiana P. Miller

These handsome, spreading evergreen trees, with elliptic to obovate leaves 4–6 cm long, often support populations of Spanish Moss and other epiphytes. The flowers of the wind-pollinated oaks are minute and classification is based on leaf and fruit characters.

Live Oak is frequent in moist woodlands and hammocks throughout Florida and much of the southeast coastal plain.

T-OA/SEE/RAY
(L 62-1-7; R 55-3-13)
FAGACEAE

RKG

274. Alder

Alnus serrulata (Ait.) Willdenow

Although the minute red stigmas of the female or pistillate flowers of Alder are rarely noticed, the pendulous, 3–6 cm long yellow brown catkins of pollen-bearing male or staminate flowers, which appear in late winter well before the leaves, are quite conspicuous.

These wind-pollinated shrubs are relatively frequent along stream margins and in moist open lowlands from northern Florida throughout the Southeast and beyond.

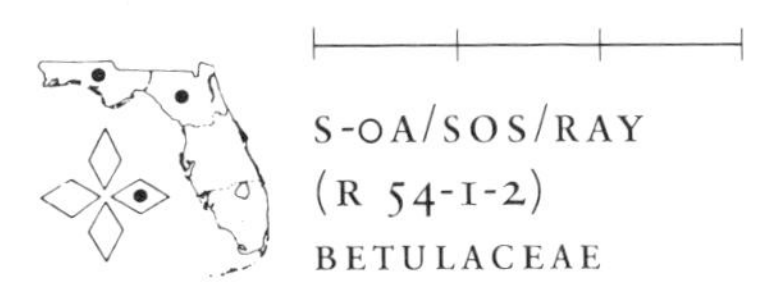

S-OA/SOS/RAY
(R 54-1-2)
BETULACEAE

BJT

275. Red Chokeberry

Aronia arbutifolia (L.) Persoon

The cluster of 1.5–2 cm white flowers with their numerous stamens and the alternate, elliptic, finely serrate leaves 4–10 cm long help identify this shrub in bloom, and the cluster of small, red, applelike fruits help identify it later in the year.

These rhizomatous shrubs often form large thickets along stream or pond margins or in bogs or moist open lowlands of northern and central Florida and much of the eastern United States.

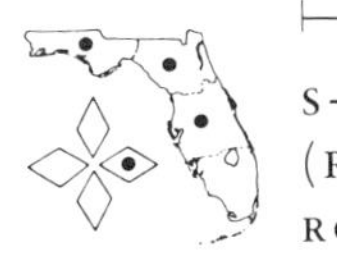

S-5A/SES/URW
(R 97-19-2)
ROSACEAE

BJT

276. Coco Plum

Chrysobalanus icaco Linnaeus

The round to obovoid edible fruits, 2–5 cm long, may be white to purple. The glossy evergreen leaves of these tropical shrubs or small trees are round to obovate, 2–8 cm long, and notched or emarginate at the tip.

Coco Plum reaches its northern limit along coastal beaches and hammocks of central Florida.

S-5A/SRE/URW
(L 94-1-1)
ROSACEAE

BJT

277. Hawthorn

Crataegus uniflora Muenchhausen

This shrub or small tree, with usually solitary flowers 1–1.5 cm across, is one of the twelve Hawthorns in northern and central Florida. (The similar *C. aestivalis* usually has five flowers in each inflorescence.)

This northern species, which ranges from New York to Missouri and Texas, is found in mixed deciduous woods and on sandhills of northern Florida, the southern limit for the species.

S-5A/SES/SRW
(R 97-20-1)
ROSACEAE

RCS

278. Mock Strawberry

Duchesnia indica (Andr.) Focke

The yellow flowers and the pulpy tasteless fruits of this creeping perennial are each about 1 cm in diameter. The trifoliate leaves resemble those of the more northern Wild Strawberry, which does not seem to occur in Florida.

This introduced Asiatic weed is often found in lawns, pastures, open woods, and along roadsides throughout much of the Southeast, but it occurs only in a few counties of northern Florida.

H-5A/TES/SRY
(R 97-2-1)
ROSACEAE

BJT

279. Gopher Apple

Licania michauxii Prance

The thick primary stems of this interesting low shrub are all underground. The slender, erect, above ground stems, seldom more than 3 or 4 dm tall, bear panicles of small flowers and evergreen, oblanceolate leaves 4–10 cm long.

A frequent, but often unnoticed, element of the sandhill flora, Gopher Apple is found throughout Florida and on the coastal plain to Mississippi and the Carolinas.

S-5A/SBE/PRW
(L 94-2-1; R 97-23-1)
ROSACEAE

BJT

280. Hog Plum

Prunus umbellata Elliott

The flowers of this shrub or small tree are 1.5–2.5 cm across; the black, somewhat fleshy, edible plums are small (only 1–1.5 cm long) and usually rather tart.

This widespread, variable, attractive woody weed often forms thick colonies along fencerows, woodland margins, and in old fields of central and northern Florida, north to the Carolinas and west to Mississippi.

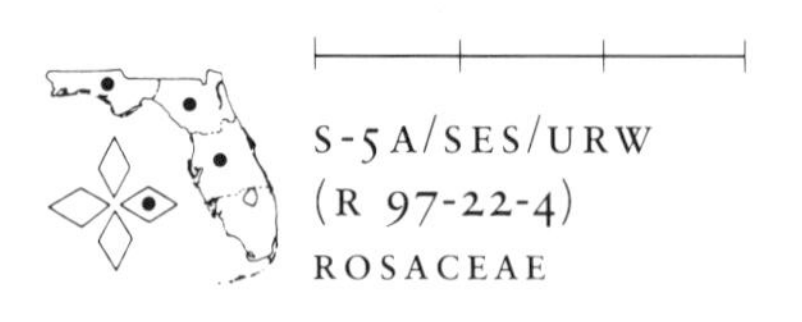

S-5A/SES/URW
(R 97-22-4)
ROSACEAE

RCS

281. Swamp Rose

Rosa palustris Marshall

This attractive, upright, rhizomatous shrub may be up to 2 meters tall. The showy but short-lived flowers are 5–8 cm in diameter. (The similar Wild Rose, *R. carolina*, occupies drier habitats and the introduced but naturalized *R. bracteata*, found into south Florida, has white petals.)

Relatively frequent along stream and pond or lake margins and in swamps or open low ground from central Florida north to Canada.

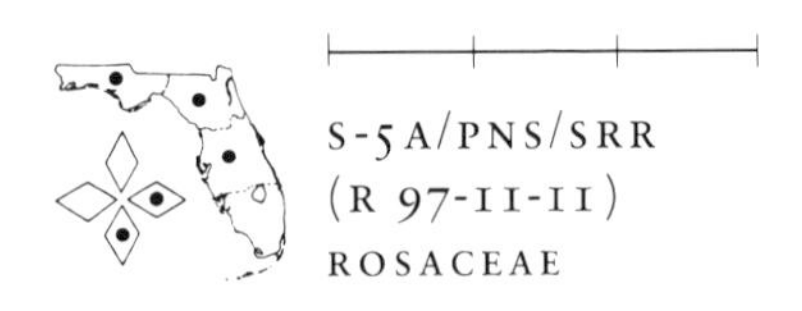

S-5A/PNS/SRR
(R 97-11-11)
ROSACEAE

BJT

282. Sand Blackberry

Rubus cuneifolius Pursh

The white, tomentose, undersurface of the leaflets quickly identify this erect native Blackberry. (The leaves of *R. argutus*, also with erect stems, are green beneath; the Dewberry, *R. trivialis*, is prostrate.)

Sand Blackberry occurs along roadsides, woodland margins, and in clearings throughout Florida and north into the Carolinas.

S-5A/TES/URW
(L 93-1-1; R 97-5-6)
ROSACEAE

CRB

283. Crab's Eye

Abrus precatorius Linnaeus

The small but brilliantly colored scarlet and black seeds of this introduced woody vine, which are only 2–3 mm in diameter, are very poisonous and may be fatal if eaten. The typical winged, zygomorphic flowers of these bean vines are reddish brown and 1–1.5 cm long.

Naturalized in roadside thickets, waste areas, and along fencerows of south and central Florida.

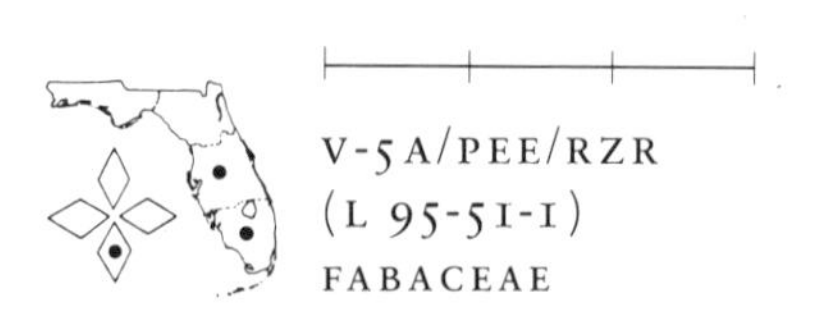

V-5A/PEE/RZR
(L 95-51-1)
FABACEAE

BJT

284. Pine Acacia

Acacia pinetorum (Small) Hermann

These 3–4 meter tall shrubs have slender flexuous branches with glabrous spines to 2 cm long and smooth, beaked legumes or fruits 3–6 cm long. (The similar *A. farnesiana*, of south and central Florida, has rounded legumes; *A. chloriophylla*, of the Keys, has no spines.)

Pine Acacia is a relatively rare endemic to the shell mounds, pinelands, and coastal strands of south Florida.

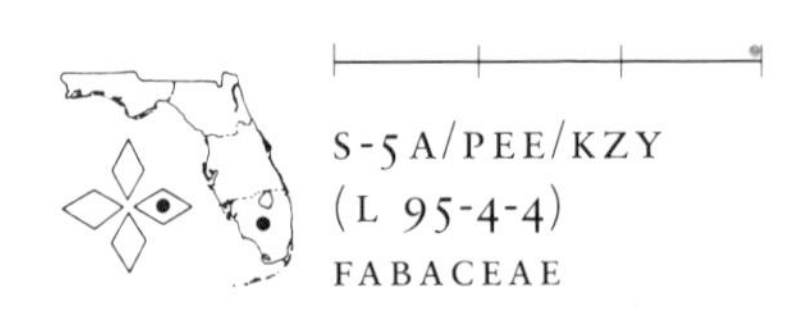

S-5A/PEE/KZY
(L 95-4-4)
FABACEAE

BJT

285. Mimosa; Silk Tree

Albizia julibrissin Durazzini

The capitate clusters of small greenish flowers are given form, size, and color by the numerous long filaments; the fragrant pink "blooms" of Mimosa are 4–8 cm in diameter and are followed by tan pods 8–15 cm long. (The leaflets of the similar and widely naturalized *A. lebbeck* are 8 mm or more wide.)

These small, spreading, introduced trees are widely planted, and are occasionally naturalized, throughout most of Florida and the Southeast.

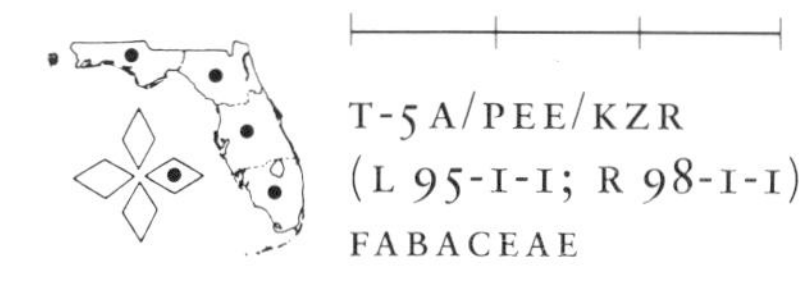

T-5A/PEE/KZR
(L 95-1-1; R 98-1-1)
FABACEAE

DNP

286. Indigo Bush

Amorpha fruticosa Linnaeus

When the dense, terminal, 1–2 dm long racemes of these bushy shrubs are in full bloom, the contrast between the golden yellow anthers and the rich purple petals of the numerous small flowers offers one of the most colorful floral combinations to be seen. (The similar Lead Plant, *A. herbacea*, is herbaceous.)

Indigo Bush is frequent along stream banks and margins of wet woods from central Florida northward over much of eastern North America.

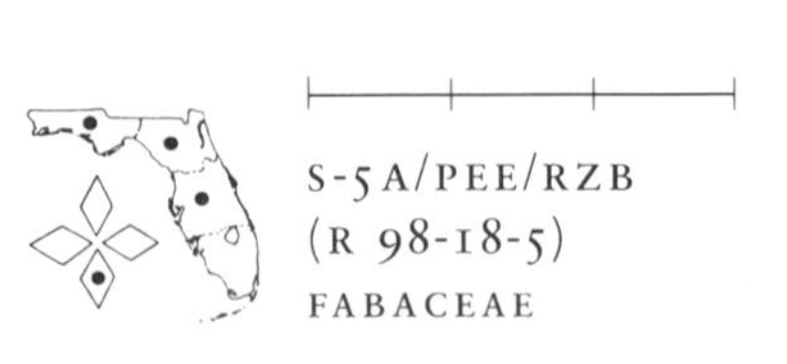

S-5A/PEE/RZB
(R 98-18-5)
FABACEAE

RCS

287. Groundnut

Apios americana Medicus

These high-climbing herbaceous vines may reach a length of 3–4 meters or more. The alternate, pinnate leaves, 1–2 dm long, have 5–7 ovate leaflets; the inflorescence of brown maroon, zygomorphic or bilaterally symmetrical, fragrant flowers can be 5–15 cm long.

Frequent in hammocks and low thickets throughout Florida and much of the central and eastern United States.

V-5A/POE/RZB
(L 95-35-1; R 98-39-1)
FABACEAE

WSJ

288. White Indigo

Baptisia alba (L.) R. Brown

The large showy flowers of this glaucous legume are 1.5–2 cm long, and the perennial plants may be a meter tall with long slender, terminal racemes 1.5–5 dm long. (*B. leucantha*, with even larger white flowers, is found south into central Florida.)

This more northern species, which reaches Tennessee and Virginia, is known in our state from only four counties of west Florida.

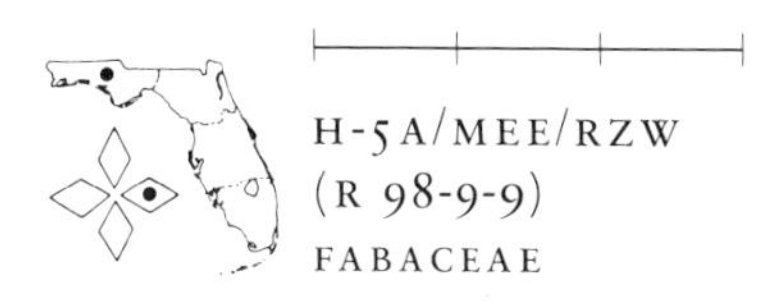

H-5A/MEE/RZW
(R 98-9-9)
FABACEAE

BJT

289. Pineland Baptisia

Baptisia lanceolata (Walt.) Elliott

The flowers of this *Baptisia* are about 2 cm long and are solitary, or two to three in number, in the leaf axils or on short terminal racemes. (*B. perfoliata*, also with yellow flowers, has simple, perfoliate leaves.)

An uncommon perennial of sandhills and open woods of central and northern Florida into Georgia and South Carolina.

H-5A/TEE/SZY
(R 98-9-6)
FABACEAE

CRB

290. Orchid Tree

Bauhinia variegata Linnaeus

The large, showy pink to white, almost regular or radially symmetrical flowers of these small trees are 5–8 cm across and are produced in late winter or early spring before the bilobed, deeply notched (actually pinnate) leaves. And the flower does indeed look somewhat like an orchid.

These horticultural plants, introduced from India, persist around old house sites or, occasionally, become naturalized in vacant lots or in thickets of south and central Florida.

T-5A/POE/SZR
FABACEAE

BJT

291. Yellow Nicker

Caesalpinia bonduc Roxburgh

The distinctive feature of these straggly, reclining, pubescent tropical shrubs is the large spiny, brown 5–7 cm long legumes, which look somewhat like the fruits of *Bixa*. (The very similar Gray Nicker, *C. crista*, may not be a biologically separate species; another of our species, *C. pauciflora*, is glabrous and has smooth legumes.)

Found in coastal hammocks and mangrove thickets from the Florida Keys to central Florida where the plants reach the northern limit of their range.

S-5A/PEE/RZY
(L 95-13-3)
FABACEAE

BJT

292. Cassia

Cassia chapmanii Isely

The three to five pairs of elliptic leaflets of the even-pinnate leaves of this *Cassia* are 2–2.5 cm long and the weakly zygomorphic flowers, characteristic of all our species of *Cassia*, are about the same size. (Coffee Senna, *C. occidentalis*, with four to five pairs of leaflets and the annual Sensitive Plant, *C. nictitans*, are both frequent in our area.)

This showy shrub of coastal pinelands of south Florida also occurs in Cuba and the Bahama islands.

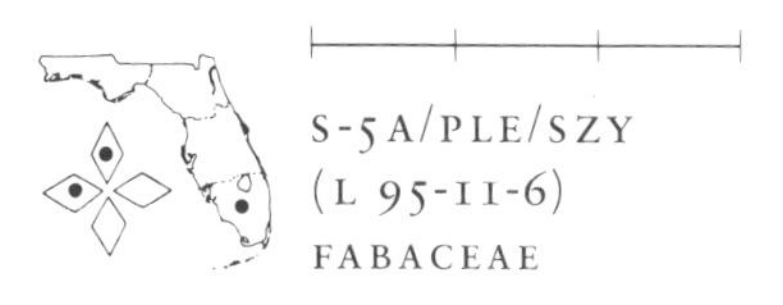

S-5A/PLE/SZY
(L 95-11-6)
FABACEAE

BJT

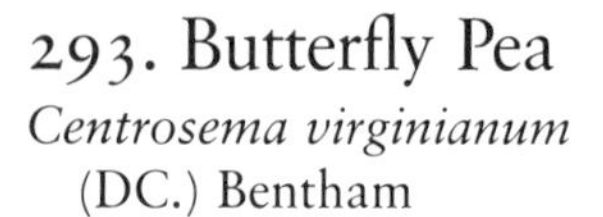

293. Butterfly Pea

Centrosema virginianum (DC.) Bentham

The standard, or large spreading upper petal, of this strongly zygomorphic flower is 2.5–3.5 cm wide. The lanceolate leaflets of the trifoliate leaves are 3–7 cm long. (The related and somewhat similar Blue Pea, *Clitoria mariana*, with five to seven leaflets, occurs in central and northern Florida.)

These herbaceous perennial vines are more or less frequent in hammocks, thickets, and open pinelands throughout Florida and much of the Southeast.

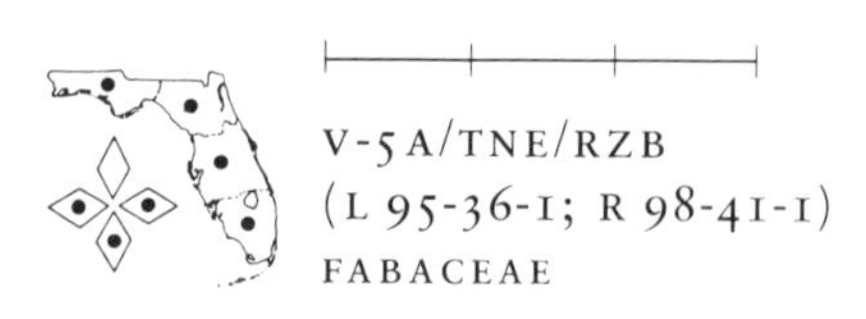

V-5A/TNE/RZB
(L 95-36-1; R 98-41-1)
FABACEAE

CRB

294. Red Bud; Judas Tree

Cercis canadensis Linnaeus

The simple, cordate leaves of this small ornamental native tree are 7–12 cm long and usually appear after the numerous small flowers, which soon produce flat pods 5–8 cm long.

Native to the eastern United States, Red Bud grows on basic soils of the rich woods of northern and central Florida.

T-5A/SCE/KZR
(R 98-4-1)
FABACEAE

BJT

295. Rattlebox

Crotalaria pallida Aiton

Each of the trifoliolate leaves of these tall herbs are 4–5 cm long, and the raceme may be 10–15 cm long. (The racemes of the showy *C. spectabilis*, which has simple, unifoliolate leaves, are 2–4 dm long.)

A frequent annual, or biennial, of roadsides, fields, and waste places in all sections of Florida and other southeastern states.

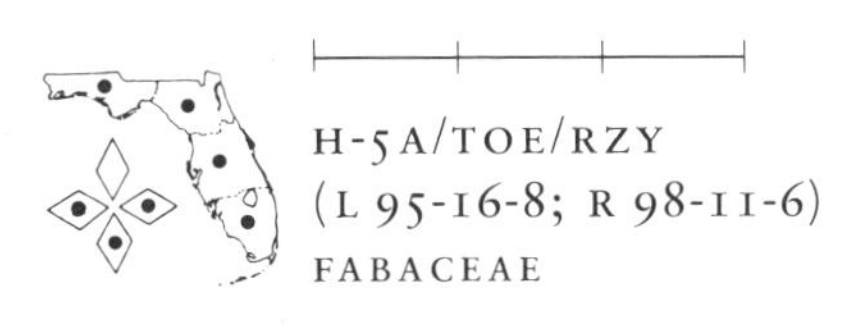

H-5A/TOE/RZY
(L 95-16-8; R 98-11-6)
FABACEAE

BJT

296. Rabbit Bells

Crotalaria rotundifolia (Walt.) Poiret

The erect stems of this small perennial are 1–4 dm long and the simple, or unifoliolate, ovate to widely elliptic leaves are 1–3 cm long. (The leaves of the related *C. purshii*, which has appressed pubescence, are linear to narrowly elliptic.)

Frequent plants of open pinelands and disturbed areas of Florida and much of the Southeast.

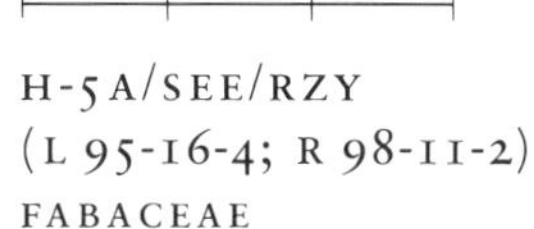

H-5A/SEE/RZY
(L 95-16-4; R 98-11-2)
FABACEAE

BJT

297. Dalea

Dalea carnea (Michx.) Poiret

The compact cylindrical spikes of numerous small pink to rose flowers are 1–2 cm long on plants 3–10 dm tall. (The spikes of *D. feayi*, also in our range, are globose.)

These erect perennials are locally frequent in open pinelands of central and northern Florida and adjacent portions of Georgia and Alabama.

H-5A/PLE/IZR
(L 95-25-0)
FABACEAE

CRB

298. Summer Farewell

Dalea pinnata (Walt.) Barneby

The compact, headlike spikes of small white flowers of these 3–12 dm tall perennials are 1–2 cm across and appear much like the heads of some members of the Aster family. The three to seven leaflets of the odd-pinnate leaves are filiform and glandular-dotted.

These herbs are native to sandhills and open sandy coastal plain woods from central Florida north to the Carolinas and west to Mississippi.

H-5A/PLE/IZW
(R 98-20-1)
FABACEAE

BCB

299. Royal Poinciana

Delonix regia (Bojer) Rafinesque

The large individual orange to crimson flowers of these flamboyant, spreading, tropical trees are 5–6 cm long, and the elongate woody pods that follow are 4–6 dm long.

A native of Africa, these ornamentals are widely planted in south Florida and have become sparingly naturalized in hammocks or have persisted around old home sites.

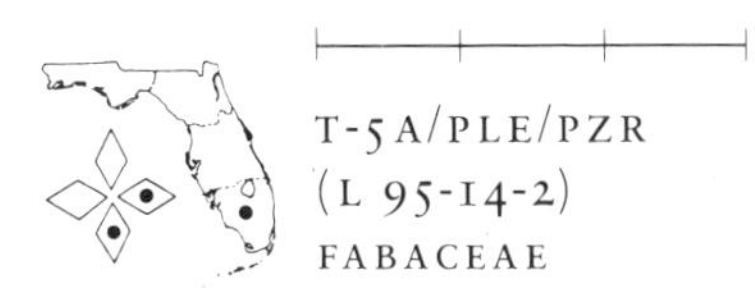

T-5A/PLE/PZR
(L 95-14-2)
FABACEAE

BJT

300. Beggar's Lice

Desmodium lineatum de Candolle

Twenty or more species of *Desmodium* occur in various parts of Florida, but few are ever noticed except when a row of their small, flat, triangular fruits are found stuck to one's clothing after an autumn walk. The banner or upper petal of the small flowers of this species is 5–6 mm wide.

These decumbent or trailing perennials are plants of open woods and sandhills primarily of northern Florida, north to the Carolinas and west to Mississippi.

H-5A/TOE/RZR
(L 95-30-0; R 98-26-8)
FABACEAE

EAH

301. Coral Bean

Erythrina herbacea Linnaeus

This shrub or small tree with brilliant flowers 2–3 cm long is our only native representative of a large group of tropical plants of the same genus. The shiny vermilion and black poisonous seeds are in black pods 7–10 cm long.

A frequent plant of the coastal woodlands and clearings of Florida, becoming less frequent to rare in Texas and the Carolinas.

S-5A/TOE/RZR
(L 95-42-1; R 98-40-1)
FABACEAE

BJT

302: Milk Pea

Galactia elliottii Nuttall

The seven to nine elliptic evergreen leaflets of this twining or prostrate vine are 2–3 cm long and are usually pubescent on both surfaces. The long racemes bear four to five flowers with banners about 1 cm across.

A common perennial of dry pinelands and white sand scrub from south Florida into Georgia and South Carolina.

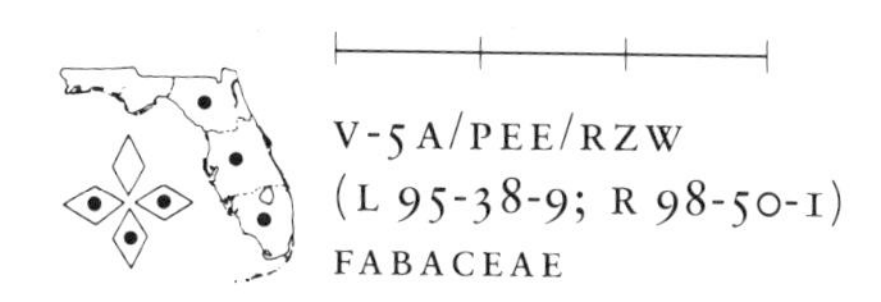

V-5A/PEE/RZW
(L 95-38-9; R 98-50-1)
FABACEAE

BJT

303. Milk Pea

Galactia pinetorum Small

The linear segments of the trifoliate leaves of this species are up to 5 cm long but only 4–6 mm wide. (The related *G. prostratum*, also endemic to south Florida, has elliptic leaflets, and the more widespread *G. regularis* has oblong or obovate leaflets.)

These perennial, trailing, herbaceous vines are endemic to the pinelands of the Everglade Keys.

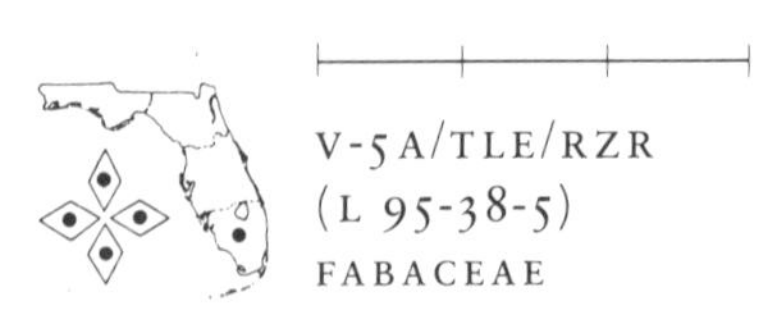

V-5A/TLE/RZR
(L 95-38-5)
FABACEAE

BJT

304. Lupine

Lupinus diffusus Nuttall

The essentially basal leaves of these silky-pubescent perennials have only a single ovate to elliptic leaflet, 5–10 cm long; such unifoliolate leaves appear to be simple. Each plant may produce numerous tall racemes of flowers with standards about 1 cm wide. (Lady Lupine, *L. villosus*, has pinkish flowers.)

An infrequent native of coastal plain pinelands, oak scrub, and sand ridges from south Florida to Mississippi and the Carolinas.

H-5B/SEE/RZB
(L 95-17-1; R 98-12-2)
FABACEAE

BJT

305. Sundial Lupine

Lupinus perennis Linnaeus

The 1.5 cm long flowers of these erect perennial herbs, with their distinctive, villous, palmately compound leaves, are usually blue but may rarely be pink or white.

This more northern Lupine is relatively rare on sandhills and in open woods across northern Florida and throughout much of the Southeast.

H-5B/MBE/RZB
(R 98-12-1)
FABACEAE

BJT

306. Cat Claw

Pithecellobium keyense Britton

The two or three pairs of leathery, elliptic to obovate leaflets that make up the bipinnate leaves of this often spiny shrub or small tree are 3–7 cm long. The black seeds are partially covered by a red fleshy tissue, called an aril, which presumably makes the seeds more attractive to birds, and thus aids in their dissemination.

These natives are endemic to the hammocks of the Keys and the southern tip of peninsular Florida.

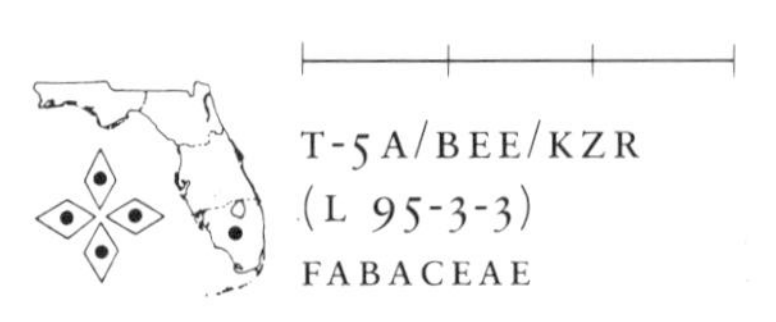

T-5A/BEE/KZR
(L 95-3-3)
FABACEAE

BJT

307. Dollar-weed

Rhynchosia reniformis (Pursh) de Candolle

The single round to slightly heart-shaped, or cordate, leaflet of this erect, 2–3 dm tall perennial with short terminal or axillary racemes, is 2–3 cm in diameter.

Frequent to common in sandy soils of open pinelands and clearings from south Florida north to the Carolinas and west to Mississippi.

H-5A/SRE/RZY
(L 95-18-1; R 98-43-1)
FABACEAE

RCS

308. Sensitive Briar

Schrankia microphylla (Dry.) J.F. Macbride

The globose heads of small flowers of this thorny prostrate vine are about 2 cm in diameter. The bipinnately compound leaves, with numerous small leaflets only 3–6 mm long, close at the slightest touch, thus the common name.

Locally frequent along roadsides, in clearings, open woods, and along woodland margins throughout Florida and much of the Southeast.

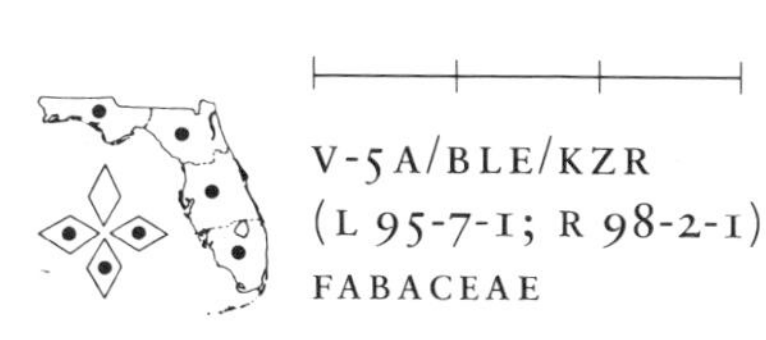

V-5A/BLE/KZR
(L 95-7-1; R 98-2-1)
FABACEAE

RCS

309. Sesban

Sesbania punicea (Cav.) Bentham

The seeds in the persistent, four-winged legumes rattle when the stalks of this 1–3 meter tall shrublike herb are shaken, thus the other common name of Rattlebox.

A widespread but not frequent introduced perennial established along roadsides, ditches, and waste places of the coastal plain from central and northern Florida to Texas and North Carolina.

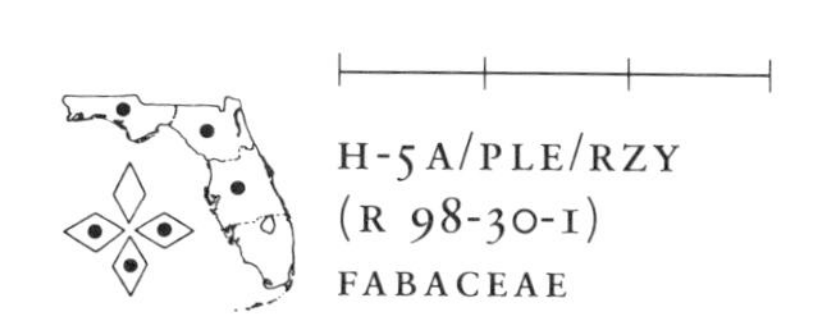

H-5A/PLE/RZY
(R 98-30-1)
FABACEAE

BJT

310. Pencil Flower

Stylosanthes hamata (L.) Taubert

The legume of this perennial herb is white pubescent. The pubescent to hirsute, elliptic leaflets of the trifoliate leaves are 2–4 cm long. (The similar and more widespread *S. biflora* occurs in central and northern Florida.)

Plants of this species are endemic to the roadsides and waste places of south Florida and possibly a few counties of central Florida.

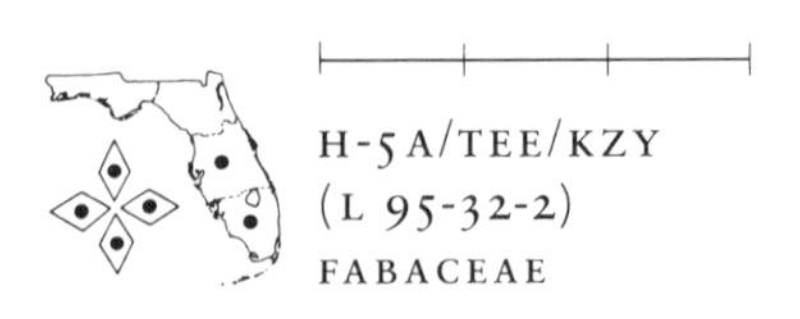

H-5A/TEE/KZY
(L 95-32-2)
FABACEAE

CRB

311. Tamarind

Tamarindus indica Linnaeus

The thick legumes of this large, spreading, introduced tree do not split open as do the legumes of many members of the bean family and are thus said to be indehiscent. The pulpy brown pods, 5–10 cm long, are often used as a spice or for seasoning.

A native of India now naturalized along the coastal strand of south Florida and the Keys.

T-5A/PEE/RZR
(L 95-10-1)
FABACEAE

BJT

312. Tephrosia

Tephrosia spicata (Walt.) Torrey & Gray

The villous, branched, spreading stems of this perennial may be 3–6 dm long. The white flowers, which turn pink or red after pollination, are solitary or two to three on a long peduncle; the rounded standard, or upper petal, is about 15 mm long. (The leaves of *T. chrysophylla*, of central and northern Florida, usually have three to seven unequal, obovate, emarginate leaflets.)

These plants are native to pinelands and dry woods of all sections of Florida and, generally, across the coastal plain of the Southeast.

H-5A/PEE/SZW
(L 95-29-1; R 98-34-3)
FABACEAE

BJT

313. Crimson Clover

Trifolium incarnatum Linnaeus

The compact showy spikes of this primarily agricultural annual, often sown as a cover crop in pecan groves, are 4–8 cm or more long. (The Hop Clovers such as *T. campestre* have small heads of yellow flowers; the common White Clover, *T. repens*, has white flowers.)

Nearly a dozen introduced species of clover, including Crimson Clover, are planted in lawns, pastures, fields, orchards, or along roadsides of all sections of Florida, and most have become established here and over much of the Southeast.

H-5A/TBS/IZR
(R 98-14-4)
FABACEAE

BJT

314. Vetch

Vicia acutifolia Elliott

The pinnate leaves of these sprawling or climbing vines are reduced to only two (or three) pairs of linear foliage leaflets, usually 2–3 cm long, and a pair of tendrils that are modified leaflets. (The six leaflets of the rare central Florida endemic, *V. ocalensis*, are usually 3–5 cm long; those of the widespread *V. floridana* are only 1–1.5 cm long.)

A perennial of hammocks, pond margins, and wet woods from south Florida into Georgia and South Carolina.

V-5A/PLE/RZB
(L 95-50-1; R 98-36-10)
FABACEAE

BJT

315. Cowpea

Vigna luteola (Jacq.) Bentham

The ovate to lanceolate, sparsely pubescent leaflets of this perennial vine may be 2–8 cm long and the pubescent, linear legumes 4–6 cm long.

This twining or prostrate cosmopolitan tropical weed is common in disturbed soil, thickets, and waste places from south Florida north to the Carolinas and along the Gulf to Mexico.

V-5A/TLE/RZY
(L 95-49-1; R 98-45-1)
FABACEAE

WSJ

316. Wisteria

Wisteria frutescens (L.) Poiret

The nine to fifteen ovate leaflets of these woody vines are 2–6 cm long, and the compact raceme may be 4–12 cm long. The linear legumes of this plant are glabrous. (The legumes of the related, cultivated, Asiatic *W. sinensis*, which is widely naturalized in the Southeast, are velvety-pubescent.)

This native perennial is infrequent on floodplains and along low woodland borders at widely scattered localities in northern Florida and on the coastal plain into Alabama and north to Virginia.

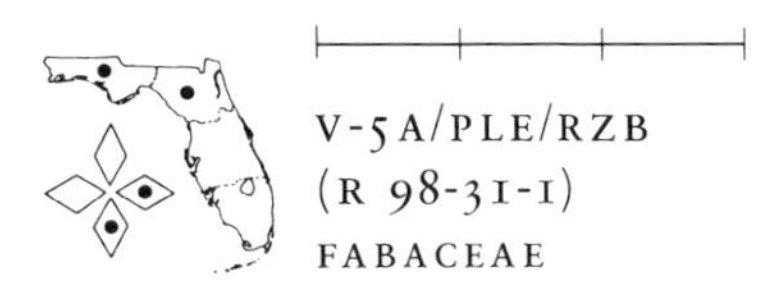

V-5A/PLE/RZB
(R 98-31-1)
FABACEAE

BJT

317. Lavender Scallops

Kalanchoe fedtschenkoi Hamet & Perrier

The fleshy, dentate, obovate leaves of the lower stems produce small plantlets in the notches of each dentation. In contrast to Life Plant, the inflorescences of this succulent perennial bear only a few flowers and the colorful petals are about twice as long as the tubular calyx.

Another introduced tropical horticultural plant that has become naturalized in waste areas and vacant lots in the normally frost-free areas of south and central Florida.

H-40/SBD/UTR
(L 90-2-0)
CRASSULACEAE

BJT

318. Life Plant

Kalanchoe pinnata (Lam.) Persoon

The thick fleshy leaves of this colorful succulent perennial are up to 1 dm long and often produce new plants at each of the small notches along the margin, even when the leaf is cut from the stem. The inflated, cylindrical calyx is about 3 cm long.

Scattered plants or dense colonies of these tender herbs, introduced from tropical Africa and now naturalized, are found in disturbed areas of lower peninsular Florida and the Keys.

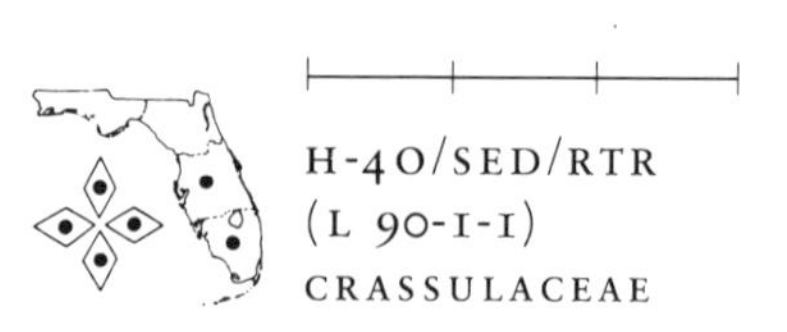

H-40/SED/RTR
(L 90-1-1)
CRASSULACEAE

BJT

319. Kalanchoe

Kalanchoe tubiflora (Harv.) Hamet

The numerous fleshy, mottled leaves of this succulent naturalized African herb, which produce small plantlets at their tips, can be from 3–15 cm long and are almost as showy as the pendulous, red to purple or brown flowers that are 2–3 cm long.

The plants, up to 1 meter tall, are not frost hardy and are found in waste places and disturbed areas of south and adjacent central Florida where they may form large colonies.

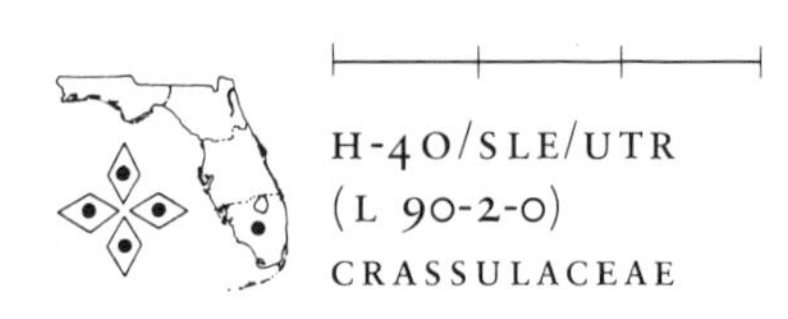

H-40/SLE/UTR
(L 90-2-0)
CRASSULACEAE

EAH

320. Oak-leaf Hydrangea

Hydrangea quercifolia Bartram

The showy white sterile flowers, which are 1–2 cm wide, presumably help attract pollinating insects to the large panicles of small fertile flowers. The twigs of these shrubs are covered with dense, reddish, tomentose pubescence. (The leaves of Mountain Hydrangea, *H. arborescens*, are not lobed.)

Native along stream margins, bluffs, and in open deciduous woods in west Florida and on into Georgia, Tennessee, and Mississippi.

S-5O/SOL/PRW
SAXIFRAGACEAE

WSJ

321. Virginia Willow

Itea virginica Linnaeus

The compact, terminal flower clusters of this native shrub are about 1 dm long. The alternate elliptic to obovate leaves are 3–8 cm long.

Widely distributed along streams and swamp margins from south Florida along the eastern coastal plain to New Jersey and through the Mississippi Valley.

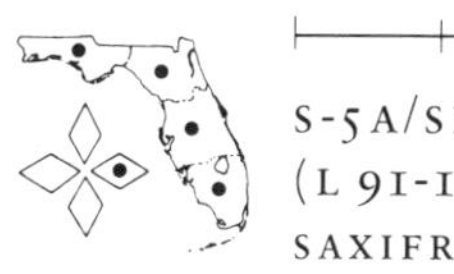

S-5A/SBS/RRW
(L 91-1-1; R 94-1-1)
SAXIFRAGACEAE

WSJ

322. Grass-of-Parnassus

Parnassia grandifolia de Candolle

The solitary, white, green-veined flowers, 2.5–4 cm in diameter, are on stalks 1–5 dm tall. The ovate basal leaves are 3–8 cm long.

This perennial herb is rare to infrequent in moist deciduous woods and along shaded stream banks from the southern Appalachians into northern Florida.

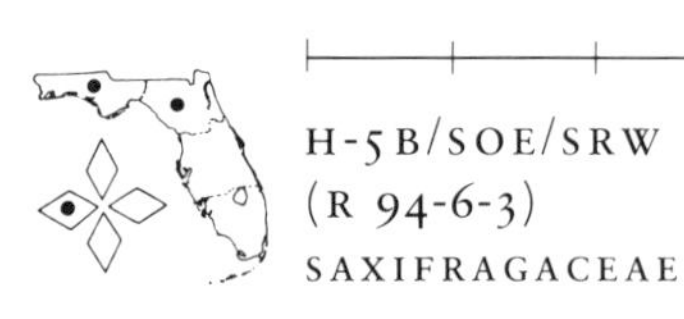

H-5B/SOE/SRW
(R 94-6-3)
SAXIFRAGACEAE

BJT

323. Sundew

Drosera capillaris Poiret

The basal rosette of colorful glandular leaves of this Sundew is usually 2–3 cm in diameter. The leafless scape, or flower stalk, is 7–15 cm tall. (The similar *D. brevifolia*, and the rare *D. intermedia* with elongate leafy stems, also occur in our area.)

This small plant is widely distributed and can be locally abundant in bogs, savannas and along wet ditches of much of Florida and the Southeast.

H-5B/SBE/RRR
(L 89-1-1; R 92-1-5)
DROSERACEAE

324. Dew Threads

Drosera filiformis Rafinesque

The basal cluster of slender filiform leaves, 1.5–3 dm long, are beset with minute but visible glandular hairs that secrete a sticky substance, which traps small insects. The glandular secretion glistens in the sun and gives these interesting plants their common name.

These cormose perennials are locally abundant in wet roadside ditches and open seepage slopes of northern Florida; less frequent on the coastal plain north and west of our area in the Southeast.

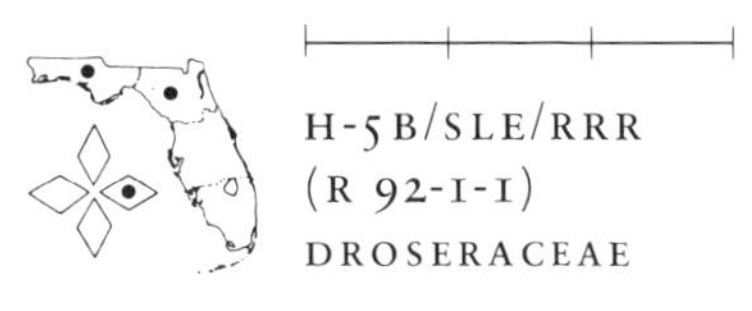

H-5B/SLE/RRR
(R 92-1-1)
DROSERACEAE

RJT

325. Buttonwood

Conocarpus erectus Linnaeus

The purplish green, conelike fruits of this tropical shrub or small tree are about 1 cm in diameter. The alternate, evergreen, obovate to elliptic leaves are 2–10 cm long. A pubescent native variety, Silver Buttonwood, is planted in south Florida as an ornamental.

These plants are found frequently on sandy coastal shores of south and central Florida.

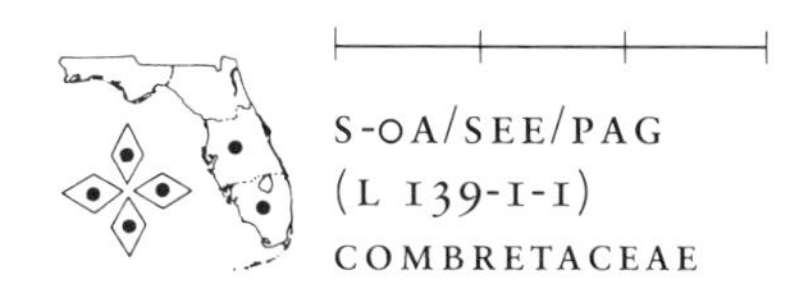

S-OA/SEE/PAG
(L 139-1-1)
COMBRETACEAE

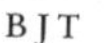

BJT

326. White Mangrove

Laguncularia racemosa (L.) Gaertner f.

The thick, opposite, coriaceous, oval leaves of White Mangrove are 4–7 cm long and have two basal glands. The drupelike fruits, red when mature, are about 1.5 cm long.

Often found with Buttonwood on the higher ground of mangrove thickets at the edge of brackish water in tidal swamps of south and central Florida.

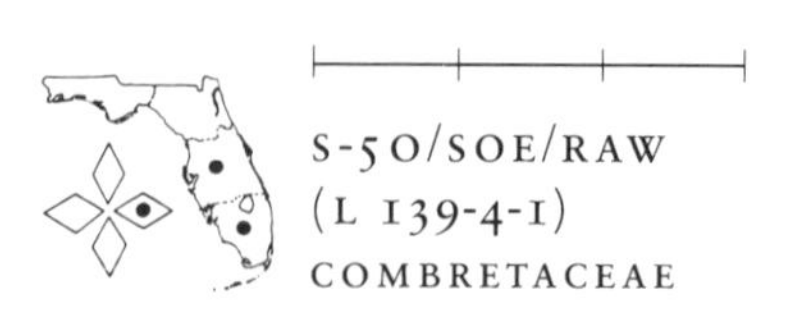

S-5O/SOE/RAW
(L 139-4-1)
COMBRETACEAE

BJT

327. Punk Tree; Bottle-brush

Melaleuca quinquenervia (Cav.) Blake

The soft, peeling white bark, the spike or "Bottle Brush" of white stamens, and the drooping branches of these introduced Australian trees make them very easy to recognize. The bright green elliptic leaves are 5–8 cm long.

Widely planted and equally widely naturalized in swamps and low woodlands of central and south Florida.

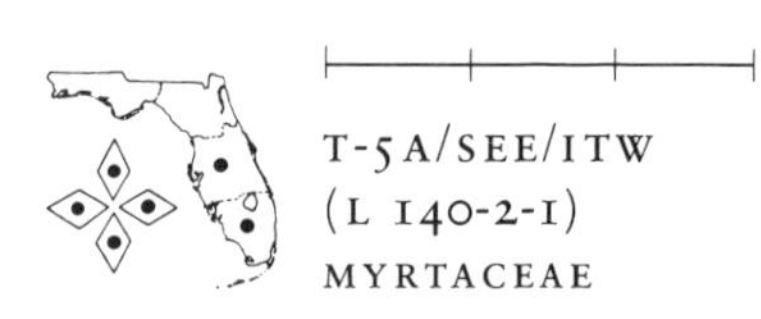

T-5A/SEE/ITW
(L 140-2-1)
MYRTACEAE

BJT

328. Nakedwood

Myrcianthes fragrans (Sw.) McVaugh

These shrubs or small trees have opposite, evergreen, obovate leaves generally 2–6 cm long and small flowers about 1 cm across. They are quite variable and difficult to classify with certainty.

A tropical species that reaches its northern limit in the hammocks of south and central Florida, where it is infrequent.

T-40/SBE/URW
(L 140-8-1)
MYRTACEAE

WSJ

329. Tall Meadow Beauty

Rhexia alifanus Walter

The glabrous stems of these perennials may be 1 meter tall. The flowers, which last only a few hours, are up to 5 cm across, and the persistent, urn-shaped hypanthium is about 1 cm long.

Native to the savannas and low pinelands of the coastal plain from central and northern Florida to Mississippi and the Carolinas.

H-40/SEE/SRR
(R 136-1-2)
MELASTOMATACEAE

WSJ

330. Yellow Rhexia *

Rhexia lutea Walter

The small but colorful plants of Yellow Rhexia, with glandular, hirsute stems, are only 3–4 dm tall and the flowers only 2–3 cm across.

An infrequent to rare perennial that grows in the coastal plain pinelands and savannas of northern Florida, west to Louisiana and north to the Carolinas.

H-40/SEE/SRY
(R 136-1-3)
MELASTOMATACEAE

BJT

331. Pale Meadow Beauty

Rhexia mariana Linnaeus

The white or rose flowers of this perennial herb, on villous-hirsute stems 3–6 dm or more tall, are 3–4 cm across. The persistent, urceolate hypanthium around the fruit is 1 cm long.

A relatively frequent native perennial in wet ditches, pinelands, and low areas throughout Florida and much of the eastern United States.

H-40/SES/SRW
(L 142-1-4; R 136-1-6)
MELASTOMATACEAE

BJT

332. Meadow Beauty
Rhexia virginica Linnaeus

The rose petals of this branching, pubescent perennial are 1–2.5 cm long, and the sessile elliptic leaves may be 3–7 cm long.

These attractive herbs, from a tuberous root, are frequent in bogs, wet pinelands, savannas, and wet ditches of northern Florida and much of the southeastern coastal plain.

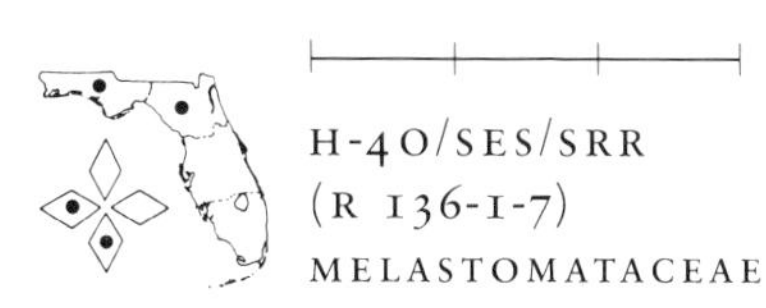

H-40/SES/SRR
(R 136-1-7)
MELASTOMATACEAE

BJT

333. Tetrazygia *
Tetrazygia bicolor (P. Mill.) Cogniau

The dark green, heavily veined, verticellate, or whorled, lanceolate leaves of this very attractive shrub are 5–10 cm long and are silvery beneath. The terminal flower clusters are about as long as the leaves; petals may number from four to six.

A rare shrub (though locally common in parts of the Everglades) of the hammocks and pinelands of our two most southern counties and the West Indies.

S-40/SNE/PRW
(L 142-2-1)
MELASTOMATACEAE

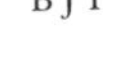

BJT

334. Gaura

Gaura angustifolia Michaux

These slender perennial herbs may be 1–2 meters tall with racemes 1–3 dm long. The delicate white flowers open in the evening and turn light pink by early the next morning when they shed their three or four small petals.

Gaura grows in open woodlands, sandy fields, and along roadsides in all parts of Florida and the outer coastal plain portions of other southeastern states.

H-4A/SES/RRW
(L 143-1-1; R 137-4-5)
ONAGRACEAE

WSJ

335. Ludwigia

Ludwigia leptocarpa (Nutt.) Hara

The showy axillary flowers of this tall, branched herb may have from five to seven petals and are 1.5–2 cm in diameter. The persistent triangular sepals crown the elongate, cylindrical capsule, which is 2–5 cm in length.

A common plant of wet ditches, low roadsides, and pond or swamp margins in all sections of Florida, west to Mississippi and north to the Carolinas.

H-5A/SLE/SRY
(L 143-3-2; R 137-1-3)
ONAGRACEAE

WSJ

336. Rattlebox

Ludwigia maritima Harper

These erect, often branched, glabrous or weakly pubescent herbs can be up to 1 meter tall; the solitary, showy flowers, 1.5–3 cm in diameter, have ovate sepals. (The sepals of the similar *L. hirtella*, also widespread in the Southeast, are narrowly triangular.)

Plants of this species occur in wet pinelands and disturbed areas throughout Florida and on to Louisiana and the Carolinas.

H-4A/SLE/SRY
(L 143-3-9; R 137-1-7)
ONAGRACEAE

BJT

337. Primrose Willow

Ludwigia peruviana (L.) Hara

These rank, herbaceous or semi-woody, branching perennials grow to 2 meters or more in height. The showy flowers may have four or five petals, each 1–3 cm long, and the leaves may vary from narrowly lanceolate to lance-ovate.

Solitary plants or large thickets of these tropical weeds are often found along swamp or pond margins and in wet ditches throughout Florida.

H-4A/SEE/SRY
(L 143-3-3)
ONAGRACEAE

BJT

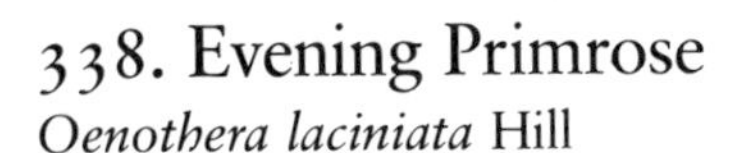

338. Evening Primrose

Oenothera laciniata Hill

The thick, pubescent, elliptic leaves, pinnately lobed to varying degree, are 3–8 cm long. The axillary flowers are 2 cm across, and the characteristic elongate capsules are 2–3 cm in length. (The leaves of the even more woody and pubescent Seaside Evening Primrose, *O. humifusa*, are entire and only 1–2 cm long.)

A common biennial weed in old fields, waste areas, and roadsides from central Florida north and west over much of the Southeast.

H-4A/SEL/SRY
(R 137-2-6)
ONAGRACEAE

RCS

339. Primrose

Oenothera speciosa Nuttall

The drooping buds of these plants open into pink or white flowers that are 5–7 cm across early in the season but get smaller as blooming progresses.

Native to the central United States, these perennials are now widely naturalized from gardens and are becoming weedy at scattered localities in Florida and other southeastern states.

H-4A/SEE/SRR
(R 137-2-7)
ONAGRACEAE

RCS

340. Butterfly Bush

Buddleja lindleyana Fortune ex Lindley

The individual tubular flowers of the compact, 1–2 dm long terminal inflorescences are 1–2 cm long and have four short corolla lobes.

These introduced horticultural shrubs have become naturalized, or have persisted after cultivation, in vacant lots, waste areas, and along stream banks in a few scattered counties of northern Florida and several other southeastern states.

S-40/SES/RTB
(R 154-5-1)
LOGANIACEAE

RCS

341. Yellow Jessamine

Gelsemium sempervirens (L.) Saint-Hilaire

These trailing or climbing woody vines have slender, wiry stems and evergreen, lanceolate leaves. The fragrant flowers, 2–3 cm long, often occur in sufficient numbers to make the plants quite showy.

Yellow Jessamine is common in thickets, clearings, and hammocks throughout Florida and the coastal plain areas of the Southeast.

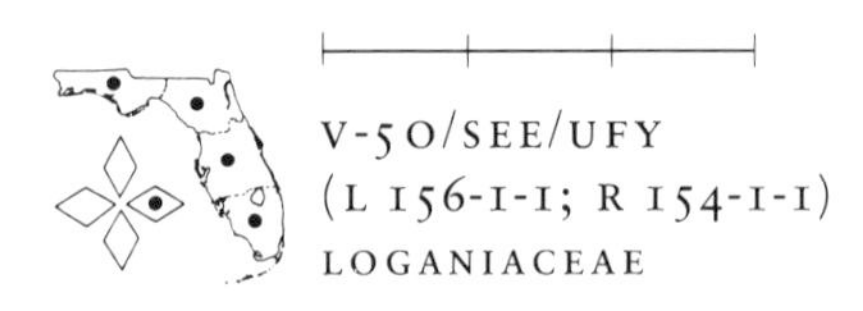

V-5 O/SEE/UFY
(L 156-1-1; R 154-1-1)
LOGANIACEAE

EAH

342. Indian Pink

Spigelia marilandica Linnaeus

Each stem of Indian Pink may bear two to twelve colorful, tubular terminal flowers 3–4.5 cm long with exserted stigmas. (*S. gentianoides* of west Florida has a pink corolla, and the tropical *S. anthelmia* of south Florida and the Keys has pale yellow flowers.)

These erect perennial herbs are rare in rich woodlands, on bluffs, and along streams at widely scattered localities in northern Florida and other southeastern states.

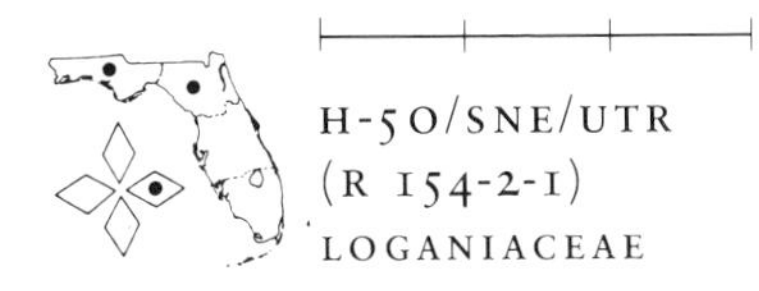

H-50/SNE/UTR
(R 154-2-1)
LOGANIACEAE

RCS

343. Button Bush

Cephalanthus occidentalis Linnaeus

The compact globose inflorescences of this shrub are 2–3 cm in diameter and are made up of many small flowers with exserted anthers. The opposite or whorled leaves are mostly 7–15 cm long.

Native to much of the eastern United States, Button Bush grows along the banks of streams, lakes, sinkhole ponds and in bay swamps and wetland prairies throughout Florida.

S-40/SOE/KTW
(L 173-1-1; R 173-2-1)
RUBIACEAE

BJT

344. Diodia

Diodia virginiana Linnaeus

The pointed, spreading petal lobes of the small, axillary, tubular flowers of this erect or decumbent perennial are 3–4 mm long.

Diodia is a common weed of low disturbed areas, roadsides, and pond or stream margins throughout Florida and the Southeast.

H-40/SNE/STW
(L 173-15-3; R 173-3-1)
RUBIACEAE

BJT

345. Beach Creeper

Ernodia littoralis Swartz

The glossy, somewhat fleshy leaves of this low, prostrate or spreading shrub are 2–4 cm long and the small, axillary, tubular flowers, which may vary from red to pink or white, are 7–10 mm long.

These tropical beach and coastal dune plants are native in south peninsular Florida, the Keys, and the islands of the West Indies.

S-40/SNE/STW
(L 173-4-1)
RUBIACEAE

BJT

346. Seven-year Apple

Genipa clusiifolia (Jacq.) Griesbach

The glossy, obovate, opposite leaves of this evergreen shrub or small tree are 5–15 cm long. The flowers, 2–3 cm across, are often in dense cymose clusters; the pulpy obovoid fruit is 7–8 cm long.

Another species apparently from the West Indies that is found in coastal areas of south Florida and the Keys.

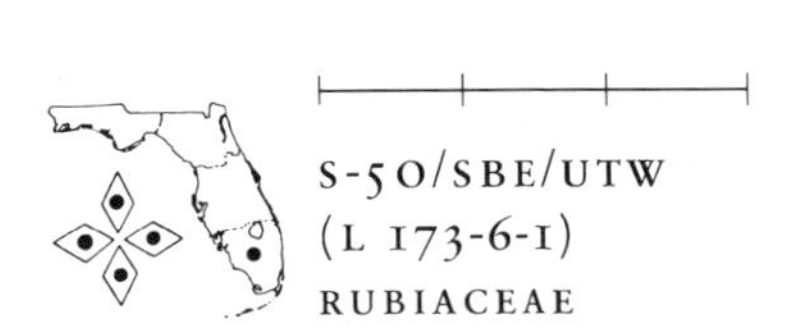

S-50/SBE/UTW
(L 173-6-1)
RUBIACEAE

BJT

347. Velvet Seed

Guettardia scabra (L.) Ventenat

The round fleshy fruits of this shrub or small tree are about 5 mm in diameter, and the tubular, 4–9 lobed corolla is 1–2 cm long. (The corolla of the related *G. elliptica* is less than 1 cm long.)

These tropical plants occur in the United States only in the hammocks and pinelands of south Floida; they are also found in the West Indies and tropical America.

S-40/SEE/UTW
(L 173-7-2)
RUBIACEAE

BJT

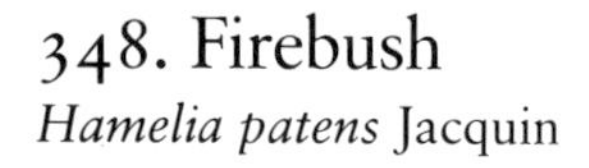

348. Firebush

Hamelia patens Jacquin

The cymose clusters of bright red tubular flowers, each 1–2 cm long, provide the common name for this showy tropical shrub, which may have either opposite or whorled leaves. The five-lobed flowers are followed by dark red berries.

A frequent plant of hammocks, coastal dunes, and shell middens of south Florida and adjacent coastal counties north to about the level of Tampa in central Florida.

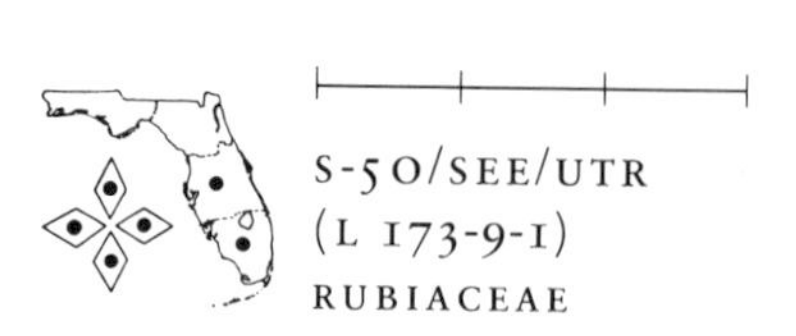

S-50/SEE/UTR
(L 173-9-1)
RUBIACEAE

WSJ

349. Innocence

Hedyotis procumbens (Walt.) Fosberg

The elliptic to round leaves of these low, creeping white-flowered relatives of the common but more northern Bluets are 5–12 mm long; the petal lobes of the tubular flowers are 4–6 mm long.

A frequent perennial of moist, open sandy areas and disturbed sites throughout Florida and along the coast to Mississippi and the Carolinas.

H-40/SEE/STW
(L 173-19-1; R 173-8-4)
RUBIACEAE

RCS

350. Partridge Berry
Mitchella repens Linnaeus

The paired tubular flowers, about 1 cm long, provide another name, Twin Flower, for these creeping woodland perennials. The colorful but tasteless fruits, about 1 cm in diameter, are apparently eaten only as a last resort by wildlife, since the berries often remain on the plant for months.

A northern plant, common in rich evergreen or deciduous forests over much of eastern North America, which may be frequent in moist woodlands of northern and central Florida.

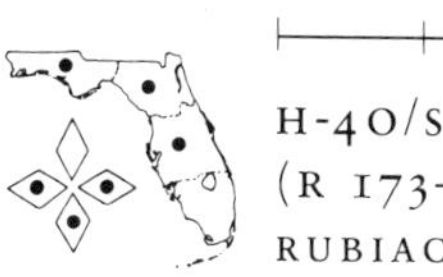

H-40/SOE/STW
(R 173-6-1)
RUBIACEAE

BJT

351. Indian Mulberry
Morinda royoc Linnaeus

The compact, capitate, or head-like, terminal inflorescences of these shrubs are 1–3 cm across and the opposite, or whorled, elliptic, accuminate leaves are 5–10 cm long. The fleshy, yellowish fruit is 2–3 cm in diameter.

An infrequent tropical shrub (or vine) primarily of coastal hammocks of south and central Florida and the West Indies.

S-40/SEE/KTW
(L 173-5-1)
RUBIACEAE

WSJ

352. Richardia
Richardia brasiliensis Gomez

The dense, strigose, or hairy, terminal inflorescence of these sprawling pubescent perennials have pubescent fruits which are 1–3 cm in diameter. (The related *R. scabra*, also of our range, has tuberculate fruits.)

This tropical weed is now naturalized in lawns, fields, roadsides, and disturbed areas throughout Florida and on the coastal plain to Texas and Virginia.

H-40/SNE/KTW
(L 173-18-1; R 173-5-1)
RUBIACEAE

BJT

353. Pineland Allamanda
Angadenia berterii (A. DC.) Miers

The large, often solitary, axillary flowers of these erect or spreading shrubs are 3–4 cm in diameter. The erect, slender, cylindrical fruits are 5–10 cm long.

This West Indian perennial occurs in the United States only in the pinelands of south Florida and the Keys.

S-50/SLE/STY
(L 158-5-1)
APOCYNACEAE

RCS

354. Blue Star
Amsonia ciliata Walter

This slender, leafy, pubescent perennial has flowers 1–2 cm across. (The related *A. tabernaemontana*, which is also in our area, is very similar but is glabrous and has wider leaves.)

Infrequent in dry pinelands of northern Florida, a few adjacent counties of central Florida, and on the coastal plain into Alabama and the Carolinas.

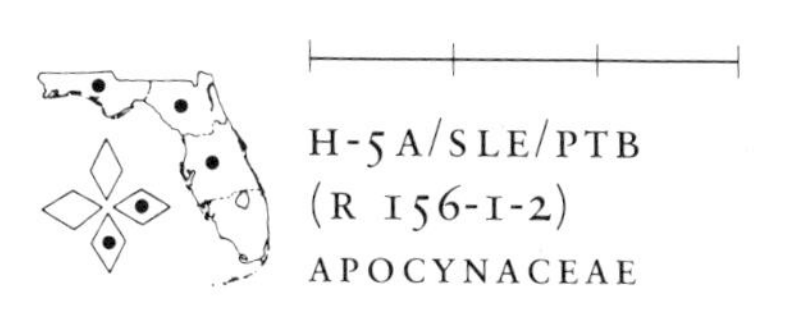

H-5A/SLE/PTB
(R 156-1-2)
APOCYNACEAE

CRB

355. Madagascar Periwinkle
Catharanthus roseus (L.) G. Don

The axillary flowers of this African native may be white, pink, or rose purple with corolla tubes 2–3 cm long. (The related *Vinca minor*, also naturalized in Florida, is evergreen and has blue flowers.)

Naturalized and becoming weedy in pinelands, waste places, and roadsides of peninsular Florida and the Keys.

H-5O/SOE/UTR
(L 158-6-1)
APOCYNACEAE

BJT

356. Rubber Vine

Echites umbellata Jacquin

This twining, semiwoody tropical vine has milky sap, opposite ovate leaves that are 5–9 cm long, and loose clusters of showy flowers 4–6 cm long.

Infrequent in the pinelands of south Florida and a few adjacent counties of central Florida; also found in the West Indies, Mexico, and South America.

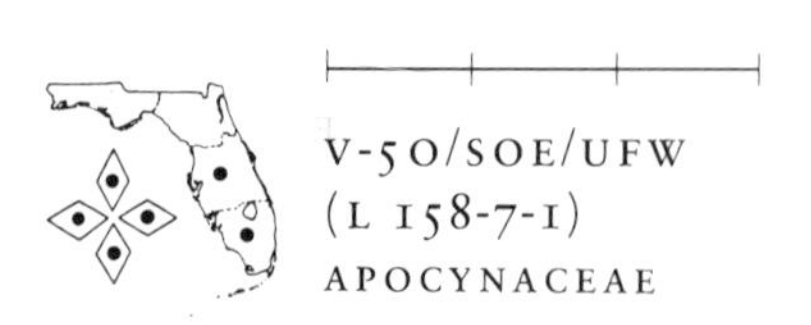

V-5O/SOE/UFW
(L 158-7-1)
APOCYNACEAE

CRB

357. Oleander

Nerium oleander Linnaeus

An introduced evergreen shrub with poisonous sap, narrow, elliptic to lanceolate, coriaceous leaves 6–12 cm long, and fragrant showy white to dark pink flowers 3–4 cm across.

Oleander is widely planted as an ornamental along roadsides and persists around old homesites in peninsular Florida.

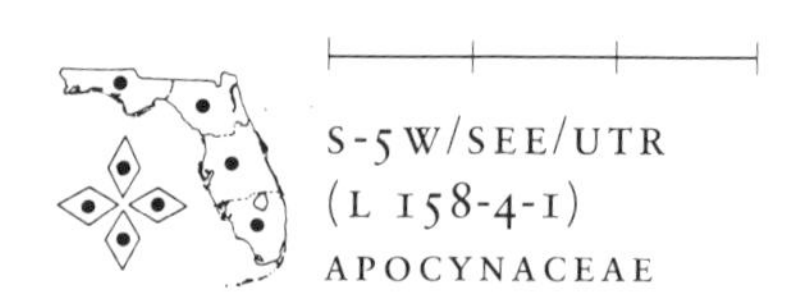

S-5W/SEE/UTR
(L 158-4-1)
APOCYNACEAE

BJT

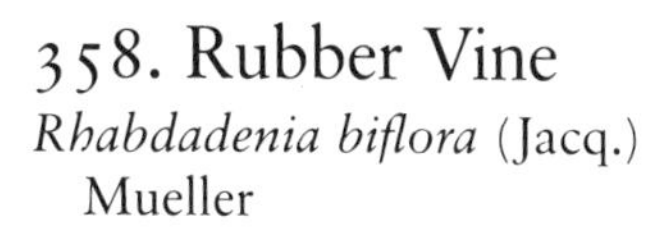

358. Rubber Vine

Rhabdadenia biflora (Jacq.) Mueller

The opposite, ovate to oblong leaves of this climbing vine are 5–9 cm long and the showy flowers 5–6 cm long.

A plant of the West Indies and tropical America that also grows in the mangrove areas and coastal hammocks of south Florida and the Keys.

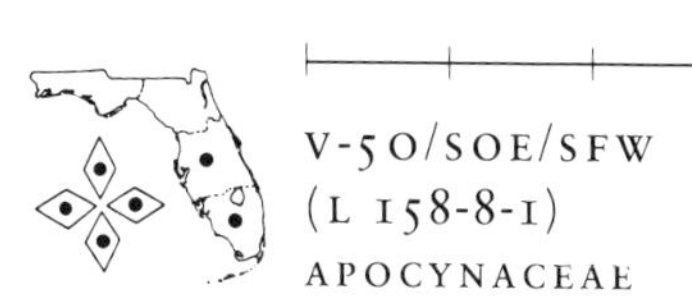

V-50/SOE/SFW
(L 158-8-1)
APOCYNACEAE

BJT

359. Wild Allamanda

Urechites lutea (L.) Britton

Equally as colorful as the true Allamanda, the smaller flowers of this glabrous or pubescent, bright green twining vine average only 5 cm across. The opposite leaves are oblong to round, 5–7 cm long.

A native of the West Indies and coastal hammocks, pinelands, and mangrove thickets of south Florida and the Keys.

V-50/SBE/STY
(L 158-9-1)
APOCYNACEAE

EAH

360. Scarlet Milkweed

Asclepias curassavica Linnaeus

The contrast between the scarlet petals and the orange to yellow hoods, which hold the copious nectar of these interesting flowers, makes this naturalized tropical introduction one of our most colorful Milkweeds.

Our only annual species of *Asclepias*, this pantropical ornamental weed has become established on sandy disturbed soils in south and central Florida.

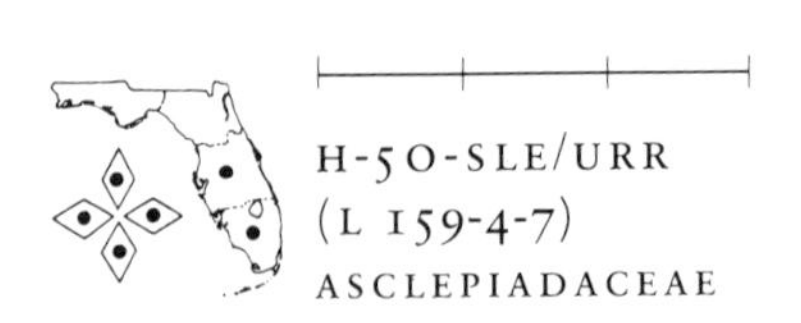

H-5O-SLE/URR
(L 159-4-7)
ASCLEPIADACEAE

BJT

361. Purple Milkweed

Asclepias humistrata Walter

The prostrate stems of this perennial are usually up to 1 meter long, turn up at the end, and have rounded, opposite leaves, 4–8 cm wide, with conspicuous purplish veins.

These native herbs occur infrequently in pinelands, sandy scrub, and on sandhills throughout Florida and on the coastal plain to Mississippi and North Carolina.

H-5O/SOE/URW
(L 159-4-4; R 157-1-11)
ASCLEPIADACEAE

CRB

362. Swamp Milkweed

Asclepias incarnata Linnaeus

The globose flower clusters of these native perennials are 5–10 cm in diameter, and the lanceolate to elliptic leaves are 6–15 cm long.

More frequent northward, this is a conspicuous but not common plant of moist open meadows and swamp or pond margins throughout Florida.

H-5O/SLE/URR
(L 159-4-8; R 157-1-1)
ASCLEPIADACEAE

DNP

363. Longleaf Milkweed

Asclepias longifolia Michaux

The small clusters of white flowers and the slender stems with numerous opposite, elongate, linear leaves 8–14 cm long help identify this low perennial herb.

Widespread but rare to frequent on the moist pinelands and prairies of Florida and on to Texas and Virginia.

H-5O/SLE/URW
(L 159-4-10; R 157-1-18)
ASCLEPIADACEAE

BJT

364. Pedicellate Milkweed

Asclepias pedicellata Walter

With slender simple stems only 2–4 dm long, this is our smallest Milkweed. The pale green flowers, however, are of average size for plants of this genus.

A rare, or easily overlooked, perennial of coastal savannas and pinelands in all parts of Florida, and up the coast to North Carolina.

H-5O/SLE/URG
(L 159-4-5; R 157-1-20)
ASCLEPIADACEAE

CRB

365. Butterfly Weed

Asclepias tuberosa Linnaeus

This native perennial differs from our other species of *Asclepias* by having alternate leaves and clear sap. Often cultivated for its brilliant flowers, which are yellow in some of our plants, Butterfly Weed may have several stems up to 1 meter long from a single root crown.

A widespread and relatively frequent herb of dry fields and roadsides more or less throughout our area and the eastern United States.

H-5A/SNE/URY
(L 159-4-3; R 157-1-4)
ASCLEPIADACEAE

BJT

366. Variegated Milkweed

Asclepias variegata Linnaeus

The two to four compact umbellate flower clusters of this erect, unbranched Milkweed are 5–10 cm or more across.

These more northern native perennials, found infrequently along the margins of rather dry upland woods, range sparingly from northern Florida over much of the central and eastern United States.

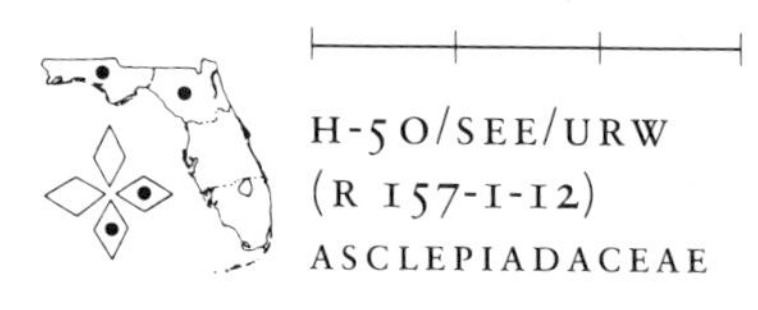

H-50/SEE/URW
(R 157-1-12)
ASCLEPIADACEAE

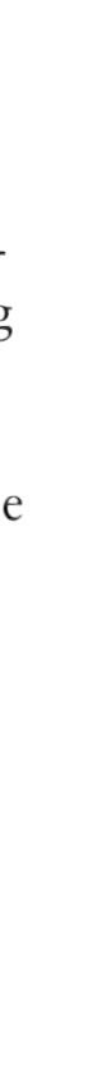

WSJ

367. Climbing Milkweed

Matelea gonocarpa (Walt.) Shinners

Only a few of the brown maroon flowers, 1–2 cm across, produce a smooth, angled seed pod, which may be 6–10 cm long. The large cordate, or heart-shaped, leaves of this twining vine are 8–12 cm long. (The rare *M. flavidula*, known only from Gadsden County, and *M. alabamensis* from Liberty County, both have yellow flowers and round, spiny pods.)

An infrequent perennial of alluvial woods or swamp forests of northern and central Florida on to Alabama and the Carolinas.

V-50/SCE/URB
(R 157-3-1)
ASCLEPIADACEAE

BJT

368. White Vine

Sarcostemma clausa (Jacq.) Roemer & Schultes

This herbaceous or semiwoody vine bears the flower clusters on stems or peduncles that are twice as long as the 6–7 cm long ovate leaves.

An infrequent introduced perennial limited to shell mounds and coastal hammocks of peninsular Florida and the Keys.

V-50/SEE/URW
(L 159-2-1)
ASCLEPIADACEAE

BJT

369. Seaside Gentian

Eustoma exaltatum (L.) Salisbury ex G. Don

Although not a true Gentian, but a member of the Gentian family, this tall annual has petals or corolla lobes about 2 cm long which may occasionally be white.

A native of the coastal sand dunes and hammocks of southern Florida, the Keys, and islands of the West Indies.

H-50/SEE/SRB
(L 157-4-1)
GENTIANACEAE

WSJ

RCS

370. Catesby Gentian

Gentiana catesbaei Walter

The stems of this native perennial are scabrous, or rough to the touch. The funnelform flowers are 3–5 cm long and occur in compact clusters.

The Catesby Gentian is a relatively rare plant of the moist sandy pinelands of northern Florida, and northward on the coastal plain to Virginia.

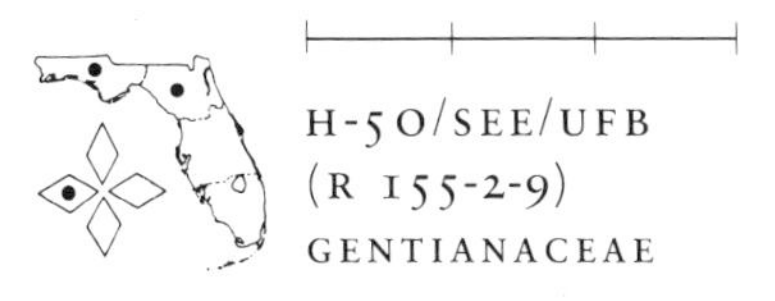

H-5O/SEE/UFB
(R 155-2-9)
GENTIANACEAE

371. Rose Pink

Sabatia angularis (L.) Pursh

The freely branched, square, and weakly winged stems of these annual herbs are 2–8 dm tall; the pink petals are 1.5–2 cm long.

Bitter-bloom, as this plant may also be called, is infrequent in the moist meadows, marshes, and fields of northern Florida and much of the Southeast.

H-5O/SOE/PRR
(R 155-1-5)
GENTIANACEAE

BJT

WSJ

372. White Sabatia

Sabatia brevifolia Rafinesque

The round, branching stems of these slender annuals are 2–5 dm tall, and the elliptic petals of the few to numerous starlike flowers are 1–1.5 cm long.

This Sabatia is frequent or common in wet pinelands of all sections of Florida, but infrequent to rare into coastal areas of Alabama and South Carolina.

H-50/SLE/URW
(L 157-5-1; R 155-1-3)
GENTIANACEAE

373. Ten-petal Sabatia

Sabatia dodecandra (L.) Britton Sterns & Poggenburg

The pink, or rarely white, corolla lobes of these rhizomatous perennials are 1–2 cm long and may vary in number from six to thirteen.

This Sabatia is relatively common in wet pinelands in parts of central Florida but generally rare elsewhere in its range through northern Florida into the coastal plain areas of Mississippi and Virginia.

H-90/SNE/PRR
(R 155-1-10)
GENTIANACEAE

RCS

374. Sabatia

Sabatia grandiflora (Gray) Small

These slender herbs may be a meter tall with linear to filiform leaves 4–10 cm long. The usually solitary, showy flower is 3–5 cm in diameter.

The Large-flowered Sabatia is endemic to the low pinelands and open grassy areas primarily of south and central Florida and the Keys, but it is also reported from northern Florida.

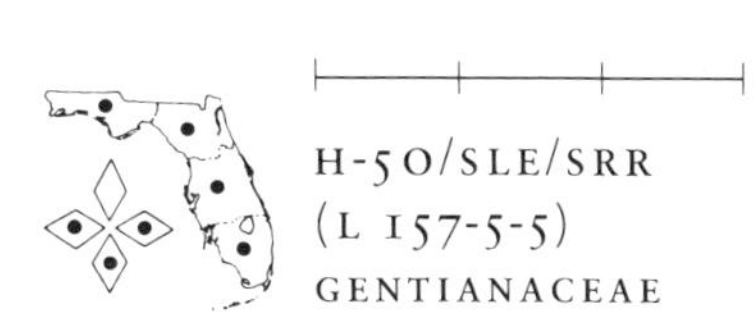

H-5O/SLE/SRR
(L 157-5-5)
GENTIANACEAE

BJT

375. Floating Hearts

Nymphoides aquatica (Walt.) Kuntze

The cluster of small flowers of this native aquatic herb appears to be borne in the notch of the heart-shaped, 5–15 cm long floating leaves. (The related *N. cordata*, of northern Florida, has mottled green and purple leaves.)

Widespread but relatively infrequent in the shallow margins of ponds and slow streams of Florida and along the coastal plain to Texas and New Jersey.

A-5B/SCE/URW
(L 157-1-1; R 155-6-2)
MENYANTHACEAE

CRB

376. Cross Vine

Bignonia capreolata Linnaeus

The colorful flowers of this high-climbing woody vine are 5–8 cm long and often occur in large axillary clusters. The semievergreen, pinnately compound leaves are reduced to two oblong to lanceolate leaflets, 6–15 cm long, with a slender tendril between them.

A frequent component of thickets, alluvial woodlands, and woodland margins throughout our area and the Southeast.

V-5O/PLE/UTY
(L 169-6-1; R 167-1-1)
BIGNONIACEAE

DNP

377. Trumpet Vine

Campsis radicans (L.) Seemann ex Bureau

Only this species of *Campsis*, with large trumpet-shaped flowers 7–10 cm long, grows in the southeastern United States; a second species is native to Southeast Asia. Also called Cow Itch Vine, this plant may be a mild skin irritant.

This woody vine, which climbs by small aerial roots, is sometimes cultivated for its ornamental value, but it is a frequent weed along fencerows, woodland margins, and in thickets from central Florida throughout much of the Southeast.

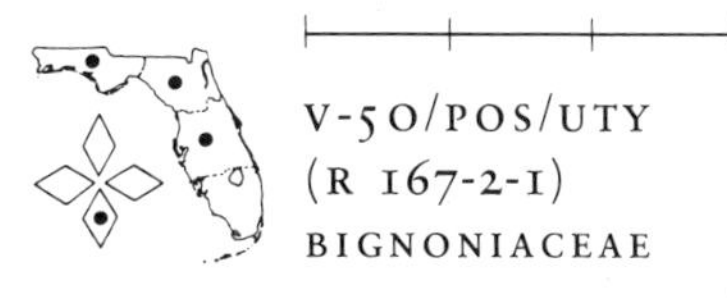

V-50/POS/UTY
(R 167-2-1)
BIGNONIACEAE

CRB

378. African Tulip Tree

Spathodea campanulata Beauvois

The colorful zygomorphic flowers of this large African tree are 5–8 cm long and are produced in compact terminal clusters.

Several exceptionally decorative, exotic tropical species of plants in the Bignonia family such as Flame Vine (*Pyrostegia venusta*), Tabebuia (*Tabebuia caraiba*), and the African Tulip Tree are planted in south Florida and may persist around old homesites or occasionally escape to thickets or waste areas.

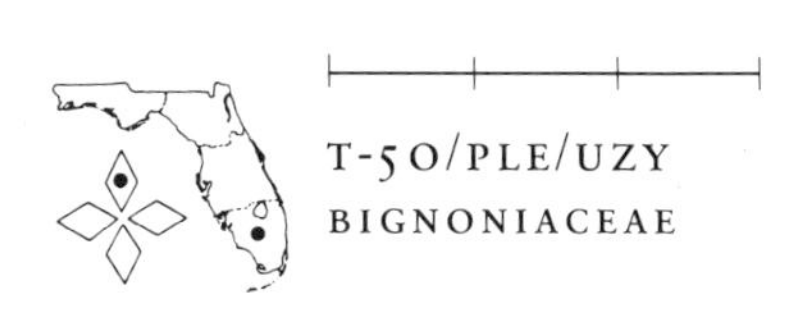

T-50/PLE/UZY
BIGNONIACEAE

BJT

379. Agalinis

Agalinis maritima (Raf.) Rafinesque

The narrow filiform leaves of this smooth, branched annual are only 2–3 cm long and may sometimes occur in bundles or fascicles. The slightly zygomorphic flowers, 12–18 mm long, have lightly fringed petals. (The stems of the similar *A. fasciculata* are scabrous.)

These semiparasitic annuals are infrequent to locally abundant in coastal marshes in all sections of Florida and along the coast to Mississippi and the Carolinas.

H-50/SLE/RTR
(L 168-12-1; R 166-25-2)
SCROPHULARIACEAE

CRB

380. False Foxglove

Agalinis purpurea (L.) Pennell

These slender, profusely branched, weakly scabrous plants may be a meter or more tall and bear a profusion of 2–4 cm long flowers in long, open racemes. (The similar *A. obtusifolia* has flowers less than 2 cm long.)

These semiparasitic annuals are common in open pinelands, meadows, woodland margins, and roadsides of Florida and much of the eastern United States.

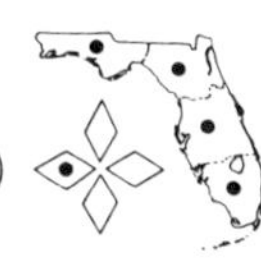

H-50/SLE/RTR
(L 168-12-6; R 166-25-3)
SCROPHULARIACEAE

BJT

381. Wild Foxglove

Aureolaria pectinata (Nutt.) Pennell

The small, glandular-pubescent, pectinate or sharply cut leaves of this annual *Aureolaria* are 2–6 cm long; the yellow axillary flowers are 3–4 cm long and about as broad.

These annuals, often semi-parasitic on the roots of Turkey Oak or other members of the black oak group, are locally frequent in dry woodlands from just below Tampa in central Florida north and west over much of the Southeast.

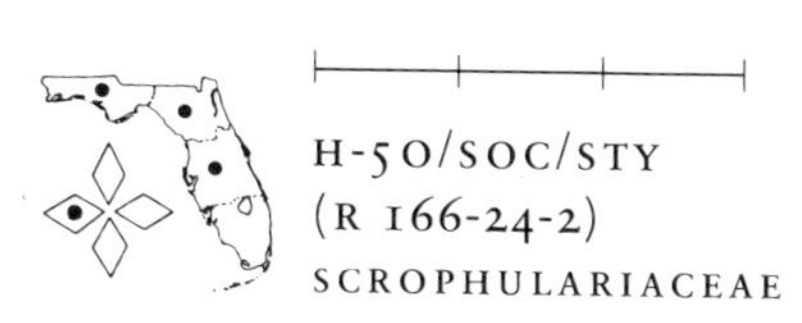

H-50/SOC/STY
(R 166-24-2)
SCROPHULARIACEAE

BJT

382. Water Hyssop

Bacopa monnieri (L.) Pennell

The solitary axillary flowers of these creeping aquatic or semiaquatic perennials are borne on slender pedicils 1–2 cm long. The small white flowers, 8–10 mm long, are only weakly zygomorphic and are often tinged with blue or pink. (Plants of the related *B. caroliniana* are very aromatic when crushed.)

Frequently forming mats on wet sand, mud flats, or along the margins of ponds or streams in all sections of Florida and on the coastal plain to Maryland and Texas.

H-50/SBE/SFW
(L 168-5-1; R 166-5-1)
SCROPHULARIACEAE

BJT

383. Buchnera

Buchnera americana Linnaeus

The erect simple stems of these hispid perennials may be 4–8 dm tall, and each bears an open spike of small, nearly symmetrical flowers with 5–8 mm long petal lobes. (The very similar *B. floridana*, with slightly smaller petals, may not be biologically distinct.)

A frequent plant of sandy roadsides, pinelands, meadows, and disturbed sites throughout Florida and, less frequently, much of the eastern United States.

H-5O/SOE/IRB
(L 168-11-2; R 166-21-1)
SCROPHULARIACEAE

EAH

384. Indian Paint Brush

Castilleja coccinea (L.) Sprengel

The brilliant inflorescence of Indian Paint Brush gets much of its texture and color from the long, lobed bracts, or modified leaves, just below each of the 2–3 cm long tubular flowers.

This colorful annual is semiparasitic on the roots of grasses in meadows, pastures, and along roadsides of much of the eastern United States but is restricted to only one or two counties in north Florida, which is the southern limit of its range.

H-4A/SOL/SZR
(R 166-28-1)
SCROPHULARIACEAE

BJT

385. Hedge Hyssop

Gratiola ramosa Walter

The small zygomorphic, or bilaterally symmetrical, flowers of these low, smooth perennial herbs are 10–15 mm long, about the length of the lanceolate leaves. (The similar *G. pilosa*, also found throughout our area, has hirsute stems and blooms later in the year.)

A frequent plant of moist open pinelands and disturbed soil in all sections of Florida and along the coastal plain to Mississippi and the Carolinas.

H-5O/SNS/SZY
(L 168-4-1; R 166-7-2)
SCROPHULARIACEAE

BJT

386. Lousewort

Pedicularis canadensis Linnaeus

The slender, tubular, two-lipped or zygomorphic flowers of these hairy perennials are about 2 cm long and are in a compact spike 2–3 cm long.

Wood Betony, first described from Canada, is a semiparasitic perennial of more northern deciduous woodlands and woodland borders and is found in only a few counties of west Florida, the southern limit for the species.

H-4A/SNL/IZY
(R 166-29-1)
SCROPHULARIACEAE

BJT

387. Beard-tongue

Penstemon australis Small

The flower stalks of Beard-tongue are usually 2–6 dm tall and may be glabrous or rather densely glandular-pubescent. The tubular, zygomorphic flowers are 15–25 mm long. (The flowers of *P. multiflorus*, which ranges to south Florida, are white.)

A frequent perennial herb of dry pinelands of central and northern Florida and on the coastal plain to Alabama and Virginia.

H-5A/SND/PZB
(R 166-14-4)
SCROPHULARIACEAE

BJT

388. Seymeria

Seymeria cassioides (Walt.) Blake

The small, essentially regular or rotate flowers, which have a glabrous calyx, are about 1 cm across and may be longer than the filiform segments of the pinnately divided leaves. (The calyx of *S. pectinata*, throughout Florida, is glandular-pubescent.)

These semiparasitic annuals are occasionally found along pineland roadsides or ditch banks and in disturbed soil from central Florida through much of the coastal Southeast.

H-5O/SPL/PRY
(R 166-23-1)
SCROPHULARIACEAE

BJT

389. Mullen

Verbascum virgatum Stokes

The attractive, nearly symmetrical flowers of this introduced biennial are about 2 cm across and are borne close to the stem on short pedicels less than 5 mm long. (The flowers of the similar *V. blattaria* are on pedicels 10–15 mm long; *V. thapsus* is woolly-pubescent.)

A rather rare introduced European weed of roadsides, old fields, and waste areas of northern Florida and a few widely scattered localities in a number of other states.

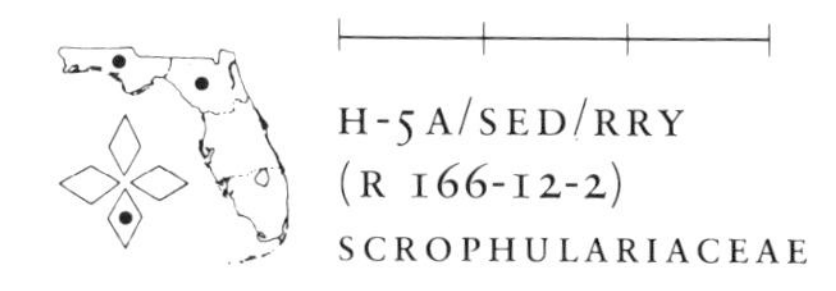

H-5A/SED/RRY
(R 166-12-2)
SCROPHULARIACEAE

BJT

390. English Plantain

Plantago lanceolata Linnaeus

The small greenish flowers of this Plantain are in compact spikes 1–8 cm long on wiry, five-angled flower stalks, or scapes, 1–5 dm long.

This introduced and thoroughly naturalized European weed is common in lawns, cleared areas, and along roadsides throughout Florida and much of the United States.

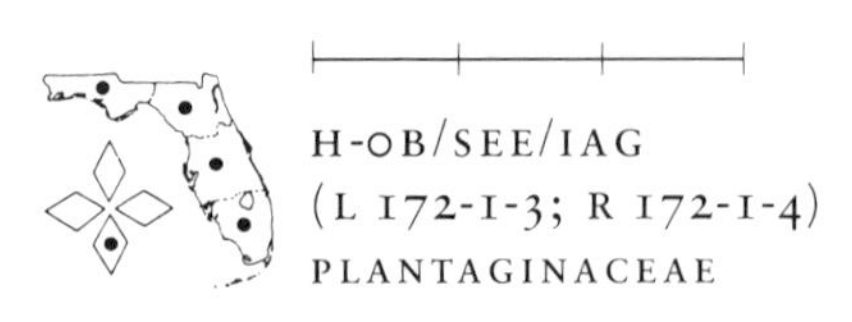

H-OB/SEE/IAG
(L 172-1-3; R 172-1-4)
PLANTAGINACEAE

WSJ

391. Southern Plantain

Plantago virginica Linnaeus

The flower spike of this pubescent Plantain measures 1 dm or more, and is equal to, or longer than, the rest of the stalk. The small bracts below each flower are not evident in this species. (The linear floral bracts of *P. aristata* are up to 1 cm longer than the flowers.)

This native annual weed is common in old fields and along roadsides in all sections of Florida and over much of the United States.

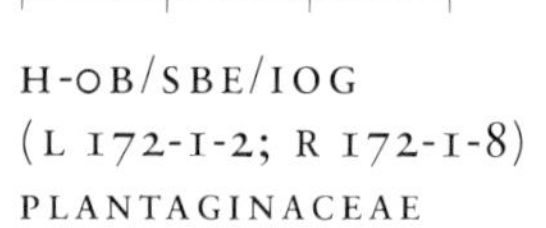

H-OB/SBE/IOG
(L 172-1-2; R 172-1-8)
PLANTAGINACEAE

RCS

392. Squawroot

Conopholis americana (L.) Wallroth

These stout, fleshy, unbranched plants, usually 1–2 dm tall, are obligate parasites on the roots of various trees and shrubs. Some plants—with self-pollinating cleistogamous, or closed, flowers—never produce above-ground stems and thus actually bloom underground.

Occasional clumps of these pale perennials are found in oak woodlands from central Florida through much of the eastern United States.

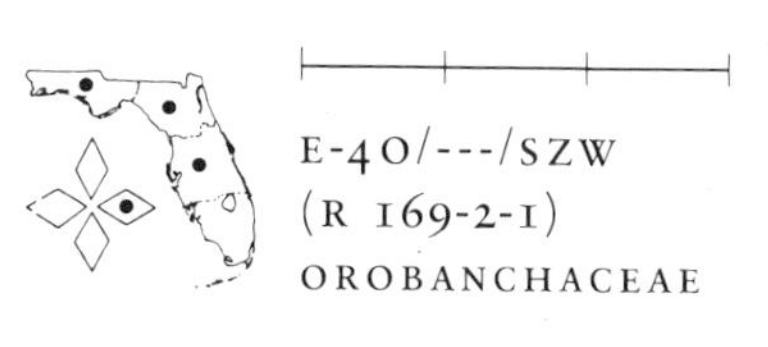

E-40/---/SZW
(R 169-2-1)
OROBANCHACEAE

MIC

393. Beech Drops

Epifagus virginiana (L.) Barton

The slender branching tan or brown stems of this parasite are usually 1–4 dm tall; the small tubular flowers are 8–12 mm long and only the cleistogamous ones are fertile.

Since these plants are obligate parasites on the roots of Beech trees, their potential range is obviously restricted to that of their host tree, which is found in northern Florida and much of the eastern United States.

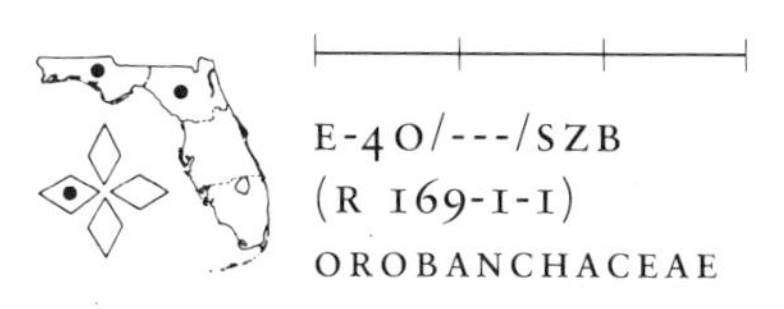

E-40/---/SZB
(R 169-1-1)
OROBANCHACEAE

WSJ

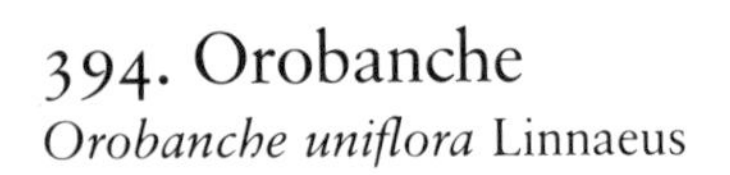

394. Orobanche

Orobanche uniflora Linnaeus

A solitary tubular flower, 12–20 mm long, is borne at the tip of each short scape of these low, bluish white, glandular-pubescent parasites.

These interesting flowering plants, parasitic on the roots of various woody plants, are rare in a few of the deciduous woodlands of northern Florida and equally rare over much of North America.

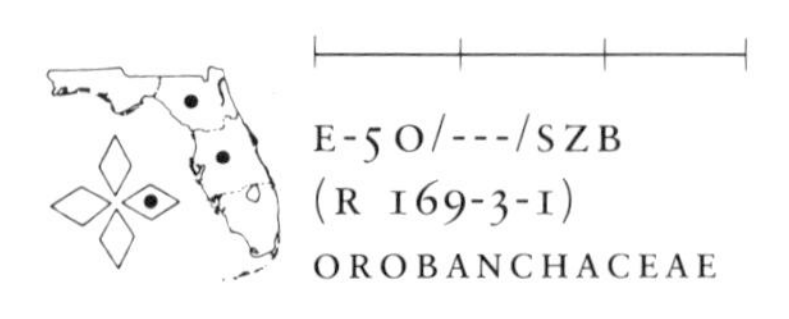

E-5O/---/SZB
(R 169-3-1)
OROBANCHACEAE

BJT

395. Blue Butterwort

Pinguicula caerulea Walter

Actually a bit more purple than blue, the solitary, weakly two-lipped, campanulate, or bell-shaped, flowers are 2–3 cm long. Small insects are trapped on the sticky leaves and provide additional nutrients for these plants.

This low perennial herb is frequent in open, moist pinelands and low roadsides from central Florida on into coastal Georgia and the Carolinas.

H-5B/SOE/SZB
(R 170-1-3)
LENTIBULARIACEAE

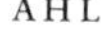

AHL

396. Yellow Butterwort
Pinguicula lutea Walter

The bright flowers of this low perennial, 2–3 cm long, are on scapes 1–3 dm long. As in other species of Butterwort, insects are trapped on the sticky leaves, which form a basal rosette 5–15 cm across.

Relatively frequent and conspicuous in wet pinelands and moist open roadsides of central Florida; less frequent on the coastal plain north to the Carolinas and west to Louisiana.

H-5B/SOE/SZY
(L 170-1-2; R 170-1-2)
LENTIBULARIACEAE

CRB

397. Flatleaf Butterwort*
Pinguicula planifolia Chapman

The deeply notched petals and the rather flat, lanceolate or elliptic reddish basal leaves, which form a rosette up to 15 cm across, are characteristic of this rare endemic. (The petals of the equally rare and restricted *P. primuliflora*, with rose flowers, and *P. ionantha*, with white flowers, are notched but not deeply lobed.)

Rare in wet soils, or even standing water, of open roadsides and pinelands of a few counties of west Florida and adjacent Alabama and Mississippi.

H-5B/SNE/SZB
LENTIBULARIACEAE

BJT

398. Small Butterwort

Pinguicula pumila Michaux

The spurred corolla of these plants are less than 2 cm long, and the rounded petal lobes are almost entire. The scape or flower stalk is often 1 dm or less tall, and the basal rosettes are only 1–2 cm across.

A frequent but easily overlooked perennial of moist pinelands of all sections of Florida, west to Texas and north to the Carolinas.

H-5B/SOE/SZB
(L 170-1-1; R 170-1-1)
LENTIBULARIACEAE

RCS

399. Horned Bladderwort

Utricularia cornuta Michaux

The spurred, strongly two-lipped flowers, on filiform stalks 1–2 dm tall, are 2–2.5 cm long. Small, modified underground leaves trap minute insects in their bladderlike traps. (The similar *U. juncea* has smaller flowers.)

These plants may grow either on very moist soil or in shallow water of ditches and pond margins throughout Florida and much of eastern North America as well as the West Indies.

H-5O/---/SZY
(L 170-2-8; R 170-2-2)
LENTIBULARIACEAE

EAH

400. Floating Bladderwort

Utricularia inflata Walter

There are two kinds of modified leaves on these plants: the radiating series of inflated floating leaves, each 1–2 dm long, which form a base for the bloom stalk and the mass of minute, submersed trap leaves.

These aquatic herbs are found in ponds, wet ditches, and canals at scattered localities from south Florida north on the coastal plain to Virginia.

A-5O/---/RZY
(L 170-2-3; R 170-2-5)
LENTIBULARIACEAE

RKP

401. Purple Bladderwort

Utricularia purpurea Walter

The submersed, whorled, compound leaves of these aquatic perennials have many small bladderlike traps that catch minute aquatic insects and provide the plants with additional nutrients. The often solitary, inflated, zygomorphic flowers are about 1 cm wide.

Found in ponds, slow streams, and wet ditches throughout Florida and, in the right habitat, sparingly over much of the eastern United States.

A-5N/---/RZB
(L 170-2-1; R 170-2-4)
LENTIBULARIACEAE

BJT

402. Crimson Dicliptera

Dicliptera assurgens (L.) Jussieu

This tropical herb has stiffly erect, branching stems 5–10 dm tall with opposite, entire, lanceolate to ovate leaves 2–12 cm long. The showy two-lipped flowers are 2–3 cm long. (A related species, *D. brachiata*, found from north Florida to Virginia, has purple or pink flowers and is not salt tolerant.)

This native perennial grows in disturbed saline sandy areas, shell mounds, and hammocks of central and south Florida and the Keys.

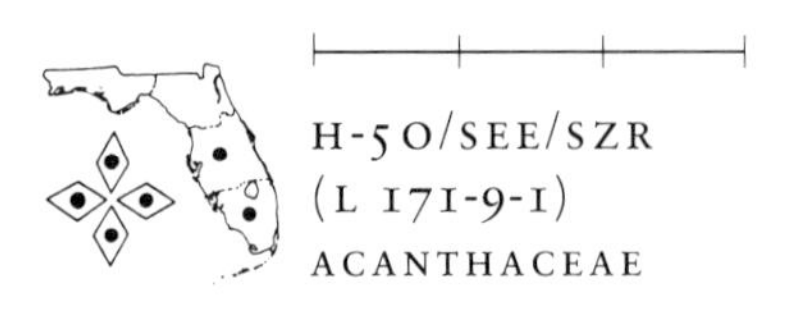

H-50/SEE/SZR
(L 171-9-1)
ACANTHACEAE

BJT

403. Twinflower

Dyschoriste oblongifolia (Michx.) Kuntze

The flowers of this small perennial herb are about 1 cm across, and are weakly two-lipped, thus differing from the closely related Ruellia.

One variety of this native perennial is endemic to the pinelands and roadsides of south Florida and the Keys; a second variety is found from central Florida to the Carolinas.

H-50/SOE/SFB
(L 171-8-1; R 171-2-1)
ACANTHACEAE

WSJ

404. Water Willow

Justicia ovata (Walt.) Lindau

Water Willow differs from both Twin Flower and Ruellia by the strongly two-lipped corolla. The opposite, sessile, ovate to elliptic or linear leaves are 3–8 cm long. (The closely related *J. crassifolia* of west Florida has flowers often 2–3 cm or more long.)

A perennial herb of moist pinelands, low woods, and coastal plain marshes from the Keys to Virginia and Texas.

H-50/SOE/SZB
(L 171-10-1; R 171-5-2)
ACANTHACEAE

BJT

405. Ruellia

Ruellia caroliniensis (J.F. Gmel.) Steudel

These variable native perennials are 3–10 dm tall with conspicuous sessile flowers, 2–5 cm across, which often last only a few hours. (The two larger, introduced species, *R. malacosperma* of south Florida and *R. brittoniana*, both have flowers on distinct stalks or peduncles.)

These widespread herbs grow in dry pinelands, deciduous woods, and woodland margins throughout Florida and the Southeast.

H-50/SLE/SFB
(L 171-7-4; R 171-3-7)
ACANTHACEAE

EAH

406. Stenandrium

Stenandrium dulce (Cav.) Nees

The weakly zygomorphic flowers of this low herb are 1.5–2 cm long, and the elliptic leaves 2–3 cm long.

This attractive little plant is endemic to the low pinelands and grassy roadsides of south and central Florida and the Keys.

H-5B/SEE/SZR
(L 171-2-1)
ACANTHACEAE

BJT

407. Red Mangrove*

Rhizophora mangle Linnaeus

The brown, cylindrical fruits, 2–3 dm long, of these evergreen shrubs are frequently seen floating in coastal waters or stranded on the beaches; the small clustered flowers, about 2 cm across, are less frequently seen. The conspicuous prop roots growing down from the branches are usually the distinguishing characteristic of Mangrove.

A common, but ecologically critical, tropical shrub or small tree of shallow marine coastlines that reaches its northern limit in south and central Florida.

S-5O/SEE/URY
(L 138-1-1)
RHIZOPHORACEAE

WSJ

408. Pepper Vine

Ampelopsis arborea (L.) Koehne

These bushy climbing vines have small greenish flowers but the clusters of red to purple berries and the large bipinnate leaves, 15–20 cm long, are the more distinctive features of this weedy perennial.

A common plant of hammocks, thickets, and fencerows from south Florida north to Virginia and west to Texas.

V-5A/BOD/URG
(L 118-1-1; R 120-3-1)
VITACEAE

BJT

409. Marine Vine

Cissus trifoliata (L.) Linnaeus

The three ovate leaflets of the trifoliolate leaves of these smooth vines are 1–3 cm long and are deeply lobed. The clusters of small greenish flowers are followed by globose, blue black berries.

Sorrel Vine, as this plant is also called, is occasional in coastal hammocks and on dunes of central and south Florida and the Keys, the West Indies, and South America.

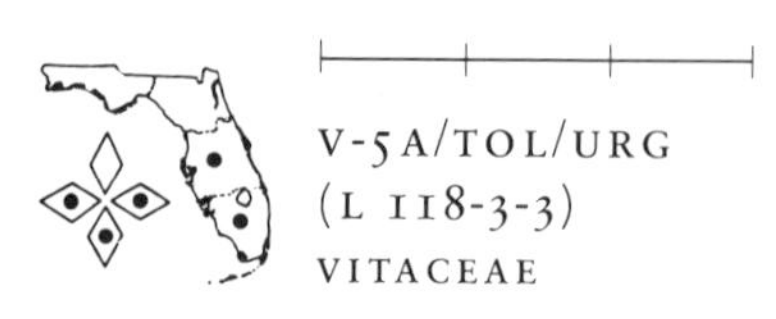

V-5A/TOL/URG
(L 118-3-3)
VITACEAE

BJT

410. Woodbine

Parthenocissus quinquefolia (L.) Planchon

The five leaflets of Virginia Creeper, as this woody vine is also called, help to distinguish this harmless plant from Poison Ivy, which has only three leaflets per leaf. The individual ovate to elliptic, attenuate or long-pointed, serrate leaflets are 6–10 cm long.

Frequent in moist hammocks, thickets, and woodland borders throughout Florida, north through much of the eastern United States, and south through the islands of the West Indies and to South America.

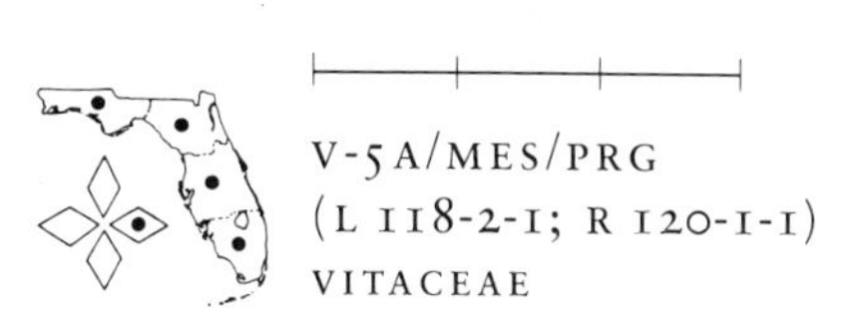

V-5A/MES/PRG
(L 118-2-1; R 120-1-1)
VITACEAE

BCB

411. Muscadine Grape

Vitis rotundifolia Michaux

The round to cordate, coarsely dentate leaves of this woody vine are 6–10 cm across. The clusters of small flowers later produce tasty, purplish round berries that may be up to 2 cm in diameter.

A common climbing or prostrate vine of open woodlands, hammocks, and dunes of Florida and the Southeast.

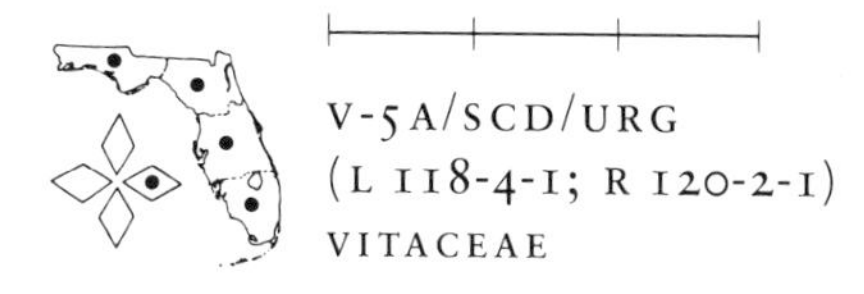

V-5A/SCD/URG
(L 118-4-1; R 120-2-1)
VITACEAE

WSJ

412. Bush Dogwood

Cornus amomum P. Miller

The opposite ovate to lanceolate leaves of these spreading shrubs are 5–10 cm long and have the prominent vein pattern typical of Dogwoods. However, the flat cymose cluster of small flowers has no large white bracts below it as in Flowering Dogwood. (The related, also more northern *C. alternifolia*, of the same limited range in Florida, has alternate leaves.)

A more northern plant that reaches its southern limit in west Florida where it has been reported in the margins of ponds, streams, or lakes in Gadsden and Jackson counties.

S-40/SOE/URW
(R 142-1-3)
CORNACEAE

WSJ

413. Flowering Dogwood

Cornus florida Linnaeus

"Florida" means "flowering," and the flowers of Flowering Dogwood are really a central cluster of small yellowish flowers surrounded by four showy bracts, 4–6 cm wide. The wood of these small trees is very hard and was once used to make wedges to split other logs.

A common tree of deciduous woodlands, thickets, and fence-rows of central and northern Florida, and much of the eastern United States, Flowering Dogwood is widely planted as an ornamental.

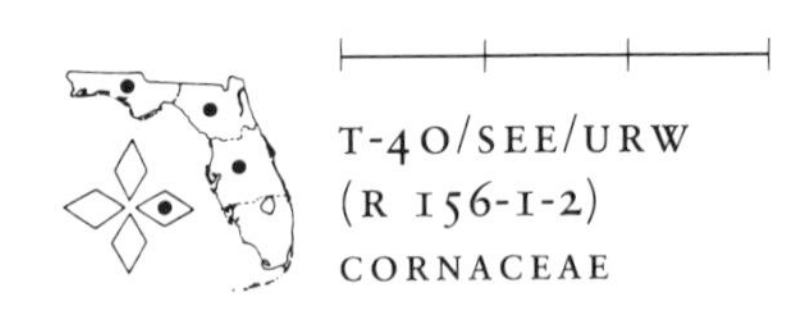

T-40/SEE/URW
(R 156-1-2)
CORNACEAE

RCS

414. Devil's Walkingstick

Aralia spinosa Linnaeus

This small tree has a thorny trunk and very large bipinnately compound leaves that may be a meter or more long and bear 150 or more elliptical leaflets 2–10 cm long. The small white flowers or dark fleshy fruits are in panicles a meter long.

This *Aralia*, sometimes planted in south Florida, is native to the southeastern United States and may occur in low woods throughout most of the state.

T-5A/BEE/PRW
(L 145-1-1; R 139-3-1)
ARALIACEAE

AHL

415. Angelica

Angelica dentata (Chapm.) Coulter & Rose

These slender perennials are up to 1 meter tall, and the flat, relatively open umbel is 5–8 cm across. The few basal leaves are deeply cut into narrow angular segments.

Only locally frequent in dry pinelands, Turkey Oak sand ridges, and clearings of west Florida and adjacent Georgia.

H-5A/SOD/URW
APIACEAE

DNP

416. Water Hemlock

Cicuta maculata Linnaeus

The large compound umbels may be up to 15 cm across, and as shown here, the terminal ones bloom before the lateral ones. These often rank biennial herbs grow from a tuberous root, have pinnate or bipinnate leaves, and are poisonous to eat. (This species is essentially inseparable from *C. mexicana*, also in our area.)

A widespread native of temperate America that is found in bogs and wet ditches at scattered localities throughout Florida.

H-5A/BLS/URW
(L 146-7-1; R 140-28-1)
APIACEAE

WSJ

417. Button Snakeroot

Eryngium yuccifolium Michaux

The round, buttonlike flowering and fruiting heads on 1-meter-tall branched stalks give this coarse perennial herb its common name; the narrow, elongate, weakly spiny, Yucca-like leaves give it part of its scientific name. (The leaves of the related *E. aquaticum* of northern Florida are elliptic and petiolate; *E. baldwinii* is prostrate and has blue flowers.)

Relatively frequent plants of moist clearings, open pinelands, and margins of deciduous woodlands throughout Florida and most regions of the Southeast.

H-5B/SLE/KRW
(L 146-1-2; R 140-4-3)
APIACEAE

CRB

418. Water Pennywort

Hydrocotyle bonariensis Lamarck

The round, shallowly notched leaves of Water Pennywort are 3–8 cm in diameter and arise from a slender, fleshy, creeping or floating stem or rhizome. The small flowers, often sterile, are usually in compound, open umbels. (The umbels of *H. umbellata* are simple; the flowers of *H. verticellata* are verticellate or whorled.)

A common aromatic perennial herb of pond and swamp margins, wet ditches, and moist sandy areas of Florida and on the coastal plain to Mississippi and the Carolinas.

H-5B/SRD/URW
(L 146-4-2; R 140-1-2)
APIACEAE

BJT

419. Water Dropwort

Oxypolis filiformis (Walt.) Britton

The hollow, slender or filiform leaves of this perennial aquatic look more like the leaves of a rush than of a member of the carrot family, but the umbellate arrangement of the flowers and the characteristic fruits clearly relate it to the Apiaceae.

A common perennial herb in wet ditches and shallow water throughout Florida and coastal areas to Louisiana and North Carolina.

H-5A/SGE/URW
(L 146-3-1; R 140-34-3)
APIACEAE

AHL

420. Purple Dropwort*

Oxypolis greenmanii Mathias & Constance

The cylindrical, jointed, hollow leaves of this aquatic perennial are even more unique than those of *O. filiformis*. The maroon flowers of this species, borne in compound umbels 8–12 cm across, are pollinated by wasps.

A rare endemic found in wet roadside ditches in only a few counties of west Florida.

H-5B/SLE/URB
APIACEAE

BJT

421. Mock Bishop Weed

Ptilimnium capillaceum (Michx.) Rafinesque

These slender, erect, branched annual herbs reach a height of up to one meter with compound umbels 2–5 cm across. The filiform leaf segments are characteristic of this species.

A relatively frequent, but ephemeral, plant of moist open areas and wet ditches of Florida and along the coastal plain to Texas and Massachusetts.

H-5A/SOC/URW
(L 146-8-1; R 140-30-4)
APIACEAE

422. Meadow Parsnip

Thaspium trifoliatum (L.) Gray

The open umbels of this perennial are 2–5 cm across. The lower leaves of both the variety shown here and the typical variety with maroon flowers are entire, but the upper leaves are divided (similar to the related *Zizia aptera* of our area).

A native of the deciduous woodlands and woodland borders of the southeastern states, these plants are also found into northern Florida.

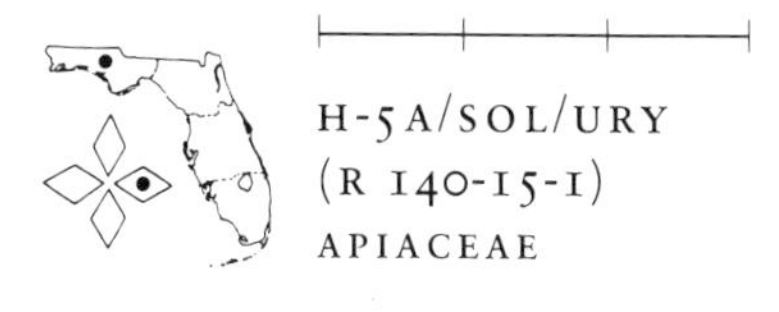

H-5A/SOL/URY
(R 140-15-1)
APIACEAE

BJT

423. Japanese Honeysuckle

Lonicera japonica Thunberg

The fragrant flowers of this woody, introduced evergreen vine are 2–3 cm long, and the black, glossy, round berries that follow are about 5 mm in diameter.

A common weed of thickets, woodlands, and waste places over much of eastern North America, Japanese Honeysuckle is found in Florida down to about Orlando.

V-5O/SEE/SZW
(R 174-2-4)
CAPRIFOLIACEAE

424. Coral Honeysuckle

Lonicera sempervirens Linnaeus

A glabrous twining vine with the opposite leaves fused around the stem just below the inflorescence. The united petals form a slender trumpet-shaped tube 3–4 cm long.

These native perennials are sometimes cultivated, but grow naturally in thickets or clearings and along fencerows generally north of Lake Okeechobee and on throughout much of the eastern United States.

V-50/SOE/UTR
(R 174-2-5)
CAPRIFOLIACEAE

BJT

425. Florida Elder

Sambucus simpsonii Rehder

The woody stems of these large shrubs have a large white pith and are easily broken. The opposite, pinnately compound leaves have 5–9 elliptic, serrate leaflets 3–8 cm long. The black fruits are edible.

These somewhat weedy shrubs, very likely indistinct from the more northern *S. canadensis*, are common on low roadsides and along fencerows and pond margins throughout the state and along the coast to Louisiana.

S-50/PES/URW
(L 174-1-1)
CAPRIFOLIACEAE

BJT

426. Possum Haw

Viburnum nudum Linnaeus

The elliptic to lanceolate leaves of these medium-sized shrubs are 5–10 cm long and have a distinct petiole that can be 1–2 cm long. The blue black fruits of this, and our other Haws, are eaten by wildlife.

A relatively frequent shrub of bogs, savannas, and low woods from the northern counties of central Florida north and west to Virginia and Mississippi.

S-5 O/SEE/URW
(R 174-5-3)
CAPRIFOLIACEAE

DNP

427. Black Haw

Viburnum obovatum Walter

The obovate leaves, 2–6 cm long, are cuneate, or tapered, to the base and are essentially sessile on the stem. The small rounded cymes, or flower clusters, of these shrubs or small trees are 2–4 dm across.

Ths is one of the more common Haws that grows in hammocks, thickets, and swamp margins in all sections of Florida and along the coast to South Carolina.

S-5 O/SBS/URW
(L 174-2-1; R 174-5-6)
CAPRIFOLIACEAE

BJT

428. Sky-flower

Hydrolea corymbosa Macbride ex Elliott

The terminal corymbose flower cluster has only a few flowers, each of which is about 2.5 cm across. The flowers are a clear blue but, unfortunately, usually photograph a lavender pink.

These weakly spiny, slender perennials are frequent in swamps and wet ditches of south and central Florida but are rare northward into coastal Georgia.

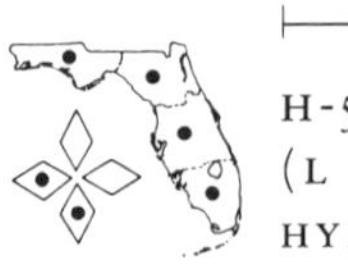

H-5A/SEE/URB
(L 162-2-1)
HYDROPHYLLACEAE

BJT

429. Geiger Tree

Cordia sebestena Linnaeus

The large terminal clusters of brilliant tubular flowers, 10–20 cm long, make this one of our most conspicuous native shrubs or small trees. The fleshy white, ovoid fruits are about 3 cm long.

The Geiger Tree, which also occurs in the West Indies, is restricted to the Everglades and Keys of south Florida, where it grows in hammocks and on dunes.

T-5A/SOE/UFR
(L 163-2-2)
BORAGINACEAE

BJT

430. Scorpion-tail

Heliotropium angiospermum Murray

These tropical annuals may be a meter tall with terminal uncoiled, scorpioid spikes of small white flowers only 1–2 mm long. The larger lanceolate-ovate leaves are 3–8 cm long and 1–4 cm wide. (The larger leaves of the similar and more widespread *H. curassavicum* are less than 1 cm wide.)

A plant of hammocks, shell mounds, roadsides, and disturbed sites in south and central Florida and the Keys.

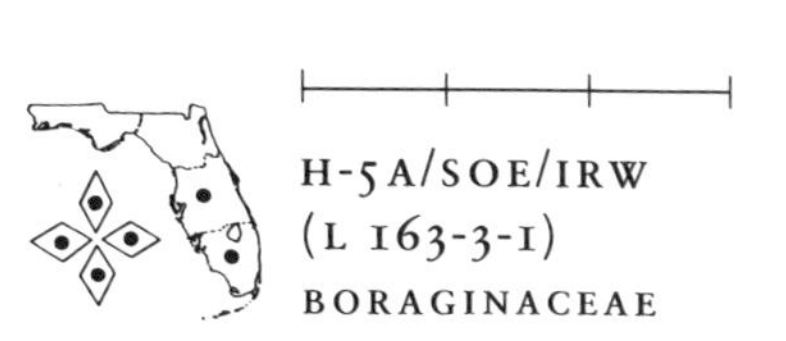

H-5A/SOE/IRW
(L 163-3-1)
BORAGINACEAE

BJT

431. Puccoon

Lithospermum caroliniense (J.F. Gmel.) MacMillan

These perennials may be up to 1 meter tall, and the hirsute stems usually have numerous branches that end in the showy coiled, then elongate, inflorescences of golden tubular flowers which are 1–2 cm long.

Infrequent to rare in dry woodlands, along sandy roadsides, and on the sandhills of northern Florida and much of the eastern United States.

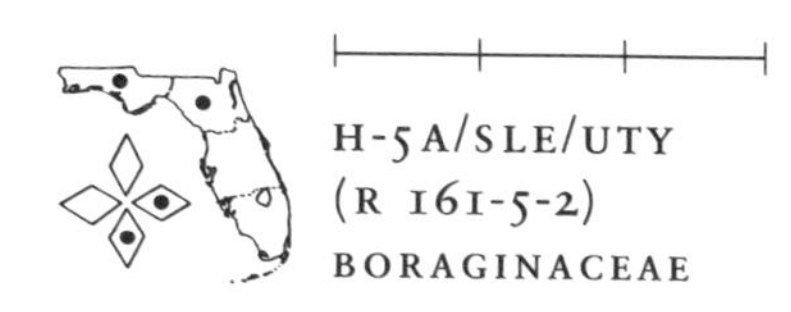

H-5A/SLE/UTY
(R 161-5-2)
BORAGINACEAE

BJT

432. Sea Lavender*

Tournefortia gnaphalodes (L.) R. Brown

This fleshy, profusely branched, maritime shrub, up to 2 meters tall, has a conspicuous silky pubescence. The crowded linear leaves are 5–10 cm long, and the small flowers are in dense, one-sided cymes.

Sea Lavender, which also occurs in the West Indies, is restricted in our area to the beaches of south and central Florida and the Keys.

S-5A/SLE/URW
(L 163-4-1)
BORAGINACEAE

BJT

433. Black Mangrove*

Avicennia germinans (L.) Linnaeus

The thick, evergreen, elliptic leaves of this tropical shrub or small tree are about 4–10 cm long and the small flowers about 1.5 cm across. The shallow horizontal roots have numerous vertical pneumatophores of aerating tissue similar in function to the "knees" of Cypress trees.

Black Mangrove is one of the more common, and more critical, broadleaf evergreens in the tidal marshes from the Keys to central Florida, and then occurs sporadically on the Gulf coast to Louisiana.

S-4O/SEE/IRW
(L 164-1-1)
VERBENACEAE

CRB

434. Beauty Berry

Callicarpa americana Linnaeus

The axillary clusters of small, light pink flowers of this scurfy-pubescent shrub are seldom noticed. However, the compact lavender pink to violet fruit clusters, 2–3 cm or more in diameter, are quite showy in late summer and give the bush its common name.

Primarily a species of the southeastern United States, Beauty Berry grows in pinelands and clearings at scattered localities more or less throughout our state.

S-50/SES/KFR
(L 165-2-1; R 162-4-1)
VERBENACEAE

RCS

435. Glorybower

Clerodendrum bungei Steudel

The large leaves of this horticultural introduction from China are 1–2 dm long and give the shrub a rather coarse appearance. The small clustered flowers have very slender corolla tubes 2–3 cm long and narrow, relatively short petals about 5–7 mm long.

An occasional escape from cultivation at widely scattered localities primarily in south and central Florida.

S-50/SOS/UTR
(L 165-3-0)
VERBENACEAE

AHL

436. Lantana

Lantana camara Linnaeus

The opposite ovate leaves of this erect or spreading shrub vary from 3–8 cm in length, and the weakly two-lipped flowers may vary from cream to yellow or pink, changing to orange or scarlet. The fruits are poison.

A native of the sandy pinelands of Florida, adjacent Georgia, and west to Texas and tropical America.

S-50/SOS/UTY
(L 165-1-2; R 162-5-2)
VERBENACEAE

WSJ

437. Capeweed

Phyla nodiflora (L.) Greene

The oblanceolate leaves of this prostrate, creeping tropical herb are 3–6 cm long, and the compact heads of small flowers are on peduncles, or flower stalks, 3–10 cm long.

Capeweed is a common plant of moist sandy roadsides, beach dunes, and disturbed areas in south and central Florida; it is relatively frequent in northern Florida, much of the Southeast, the West Indies, and tropical America.

H-50/SBS/KFB
(L 165-10-3; R 162-2-2)
VERBENACEAE

BJT

438. Blue Porterweed

Stachytarpheta jamaicensis (L.) Vahl

The slender, compact, quill-like spikes, 10–15 cm long, with the progressive ring of small blooming flowers are characteristic of these shrubby, glabrous tropical herbs. The opposite punctate, or gland-dotted, leaves are lanceolate to ovate and 2–8 cm long.

This pantropical perennial reaches its northern limit on the coastal dunes, shell mounds, and in disturbed areas of south and central Florida up to near the Tampa area.

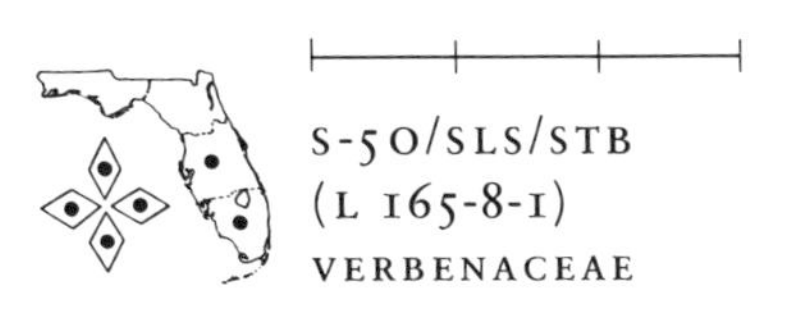

S-50/SLS/STB
(L 165-8-1)
VERBENACEAE

WSJ

439. Verbena

Verbena brasiliensis Vellozo

The branching, sharply angled stems of these scabrous perennials may be 1–2 meters tall, and the numerous terminal spikes may be 0.5–4 cm long. (The central stem leaves of the similar *V. bonariensis* have basal lobes that clasp the stem.)

This weedy tropical introduction is more or less frequent throughout our area and on the coastal plain into Mississippi and the Carolinas.

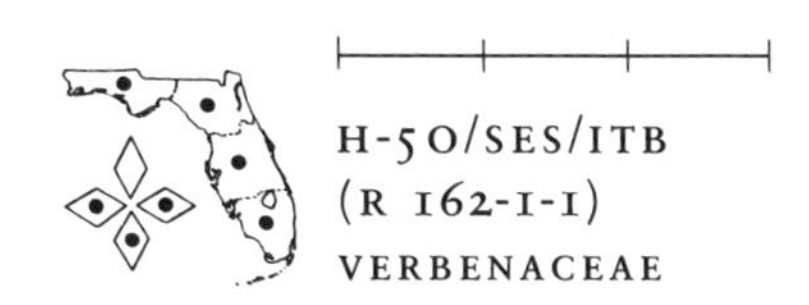

H-50/SES/ITB
(R 162-1-1)
VERBENACEAE

BJT

440. Moss Verbena

Verbena tenuisecta Briquet

The stems of these spreading or prostrate perennials root at the nodes, and a single plant may form a clump a meter in diameter. The small ovate or triangular leaves, 1–3 cm long, are divided into many linear segments.

A weedy but colorful tropical introduction, this Verbena is often abundant on roadsides and in clearings and waste areas from south Florida into the coastal plain of the Carolinas and Texas.

H-50/SOL/ITR
(L 165-9-3; R 162-1-16)
VERBENACEAE

CRB

441. Conradina

Conradina canescens (Torr. & Gray) Gray

These low canescent, or finely pubescent, shrubs have numerous, stiffly erect branches, small revolute, or rolled, leaves 1 cm or less long, and small clusters of strongly two-lipped zygomorphic flowers.

Conradina, endemic to only a small area of west Florida and adjacent Alabama, is only locally frequent in open pinelands and clearings.

S-50/SLE/UZB
LAMIACEAE

RBH

442. Scrub Balm *

Dicerandra frutescens Shinners

These low, aromatic, many-branched herbs show the need for continued, detailed study of Florida's varied flora: this species was discovered in 1962 and two additional species of *Dicerandra* have been described since. The strongly lobed zygomorphic flowers are about 1.5 cm long. (The characteristic hornlike anther appendages of *Dicerandra* are clearly shown in the inset of *D. cornutissima* which was only discovered in 1980!)

These rare herbs are endemic to the sandy scrub of a single county in the southern part of central Florida.

H-5O/SLE/RZW

LAMIACEAE

MIC

443. Dicerandra

Dicerandra linearifolia (Ell.) Bentham

These low attractive herbs are 2–4 dm tall, have linear leaves 1–3 cm long, and showy flowers about 1.5 cm long.

This species is found in open pinelands and sandhills of northern Florida and a few adjacent counties of Alabama and Georgia.

H-5O/SLE/PZR

LAMIACEAE

BJT

444. Musky Mint

Hyptis alata (Raf.) Shinners

The compact, rounded flower clusters of this rather rank perennial herb are 1.5–2.5 cm in diameter. The square stem, a characteristic of all members of the mint family, is freely branched and may be 1–2 meters tall. (The flowers of *H. mutabilis* are in a spike; those of *H. verticellata* in axillary clusters.)

A frequent weed of moist ditches and low clearings throughout Florida and along the coast to Texas and the Carolinas.

H-5O/SNS/KZW
(L 166-8-1; R 164-4-1)
LAMIACEAE

DNP

445. Lion's Ear

Leonotis nepetaefolia (L.) Aiton f.

The two or three globose flower clusters, 3–6 cm in diameter, spaced along the flowering stem and the slender orange flowers of this robust, freely branched tropical annual make it one of our most distinctive mints.

An occasional weed in thickets, barn lots, and pastures throughout Florida and other southeastern states.

H-5O/SOS/KZY
(L 166-5-1; R 164-17-1)
LAMIACEAE

RCS

446. Dotted Horsemint
Monarda punctata Linnaeus

A series of pale lavender- to cream-colored bracts just below the compact whorl of spotted, 1.5–2 cm long, yellow flowers presumably makes the infloresence of this colorful mint more attractive to insect pollinators.

Primarily a species of the Atlantic and Gulf coastal plains, this native perennial is relatively common in open sandy areas generally north of Lake Okeechobee.

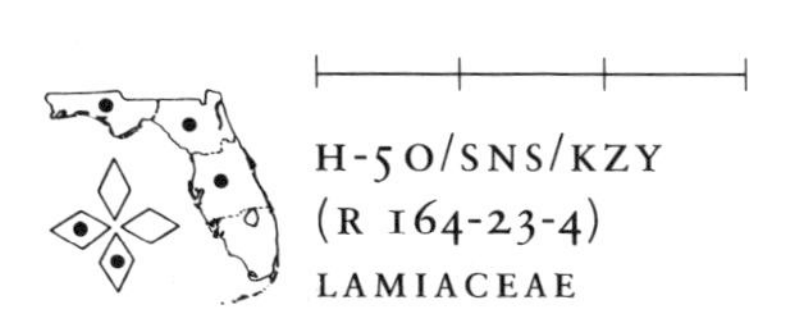

H-50/SNS/KZY
(R 164-23-4)
LAMIACEAE

BJT

447. Pennyroyal
Piloblephis rigida (Bart.) Rafinesque

The diffusely branched stems of these erect or spreading herbs have numerous, small, close-set, linear or narrowly lanceolate leaves about 1 cm long. The compact terminal raceme may be up to 6 cm long.

This aromatic plant is endemic to the dry pinelands and oak scrub of central and south Florida.

H-50/SLE/RZB
(L 166-9-1)
LAMIACEAE

BJT

448. Blue Sage

Salvia azurea Michaux & Lamarck

At the time of flowering the basal leaves of these perennials have disappeared, leaving only the linear to lanceolate stem leaves which may be 4–8 cm long. The rich blue flowers are about 1.5 cm long.

Rare or infrequent on sandhills and open pinelands of central and northern Florida and in dry open woods of other southeastern states and beyond.

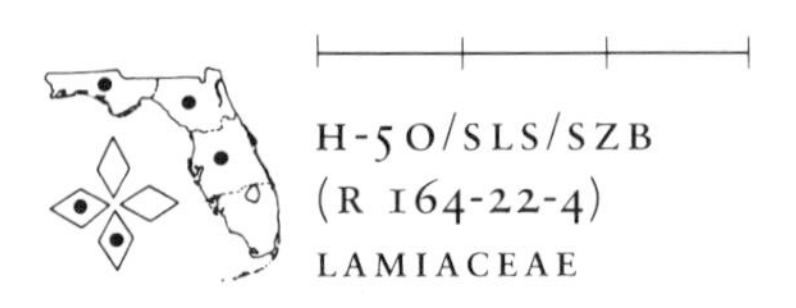

H-50/SLS/SZB
(R 164-22-4)
LAMIACEAE

WSJ

449. Tropical Sage

Salvia coccinea Jussieu ex Murray

The basal leaves of these tropical annuals are not present at flowering time. The cauline, or stem, leaves are ovate or cordate and 1–6 cm long; the strongly two-lipped scarlet flowers are 2.5–3 cm long.

A relatively frequent herb of open disturbed sites in all sections of Florida and on to Texas and South Carolina, the West Indies, and tropical America.

H-50/SOS/RZR
(L 16-6-2; R 164-22-2)
LAMIACEAE

BJT

450. Lyre-leaf Sage

Salvia lyrata Linnaeus

The scape, or flower stalk, of these weedy annuals is 3–8 dm tall and arises from a basal rosette of deeply pinnately cut, or lobed, elliptic leaves 5–15 cm long. The strongly two-lipped flowers are 1–3 cm long.

A common weed of lawns, pastures, open woods, and waste places throughout Florida and much of the eastern United States.

H-5B/SEL/SZB
(L 166-6-1; R 164-22-1)
LAMIACEAE

BJT

451. Skullcap

Scutellaria multiglandulosa (Kearney) Small ex Harper

These low, 1–3 dm tall plants are glandular-pubescent, and the leaves vary from ovate to elliptic to linear up the stem. The 2-cm-long flowers are sometimes white.

An infrequent plant of dry pinelands and scrub of northern Florida and adjacent Georgia.

H-5O/SEE/RZB
(R 164-5-5)
LAMIACEAE

BJT

452. Hedge-nettle

Stachys floridana Shuttleworth

The slender rhizomes of these perennials are terminated by fleshy, white, segmented tubers by which the plants reproduce vegetatively. The strongly two-lipped flowers, about 1 cm long, are in open whorls on the flower stalk.

Somewhat of a weed in waste areas, on roadsides, and in gardens of central and northern Florida and parts of Georgia and the Carolinas.

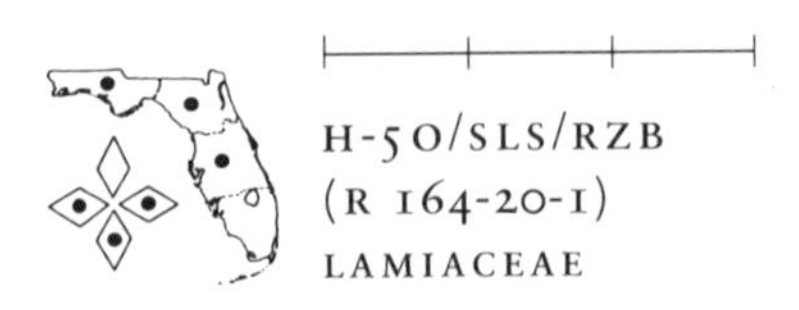

H-50/SLS/RZB
(R 164-20-1)
LAMIACEAE

RCS

453. Blue Curls

Trichostema dichotomum Linnaeus

The long curled filaments, twice as long as the small 5–10 mm long corolla, give this plant its common name. The stems of this bushy annual are covered with glandular hairs.

A frequent herb of dry woodlands and old fields in all sections of Florida, the Southeast, and beyond.

H-50/SOE/SZB
(L 166-2-2; R 164-1-2)
LAMIACEAE

RCS

454. Chamomile

Anthemis cotula Linnaeus

This strong-scented, decumbent or spreading annual, which may be 1–6 dm tall, is also called Stinking Daisy. Its scent distinguishes it from the common but more northern Ox-eye Daisy, which does not appear to grow in our area. Chamomile's conspicuous white ray flowers are 5–10 mm long.

This introduced European weed is becoming established along roadsides and in waste areas at widely scattered localities in Florida and throughout much of the United States.

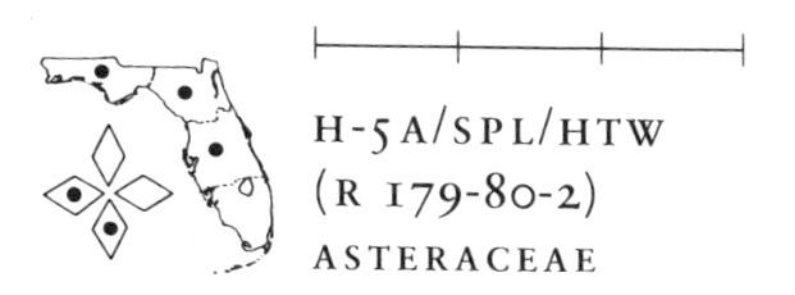

H-5A/SPL/HTW
(R 179-80-2)
ASTERACEAE

WSJ

455. Frost Aster

Aster pilosus Willdenow

A single plant of this erect, often profusely branched perennial, from 0.5–1.5 meters tall, may form a large clump 1 meter in diameter. The numerous flower heads each have fifteen to thirty-five "petals" or linear white ray flowers 5–10 mm long.

Frost Aster is infrequent in dry open woods and clearings of west Florida at the southern limit of its range, which includes much of central and eastern North America.

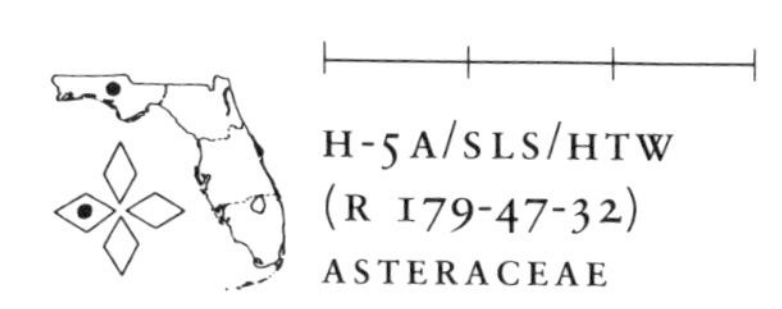

H-5A/SLS/HTW
(R 179-47-32)
ASTERACEAE

BJT

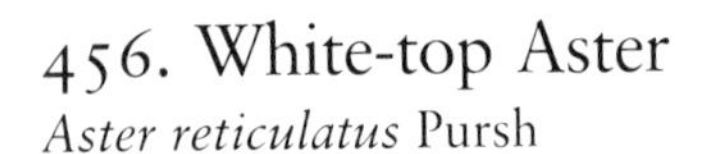

456. White-top Aster

Aster reticulatus Pursh

These downy-pubescent perennials have numerous clustered stems 4–9 dm tall and elliptic, sessile stem leaves 3–8 cm long. Each flower head has eight to twenty white rays 10–15 mm long. (Another widespread, tall perennial Aster, *A. dumosus*, has blue rays and blooms later in the season.)

A relatively common Aster of low pinelands throughout Florida and on the coastal plain to South Carolina.

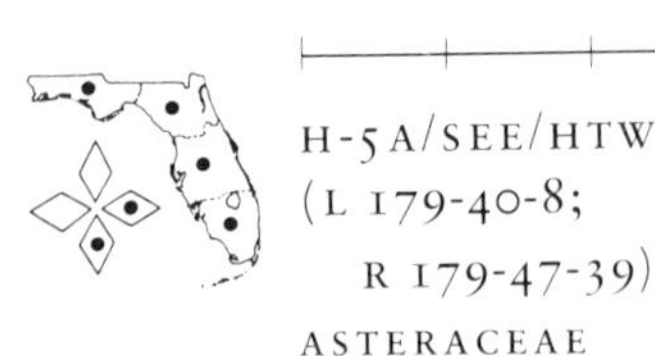

H-5A/SEE/HTW
(L 179-40-8;
R 179-47-39)
ASTERACEAE

BJT

457. Groundsel Tree

Baccharis glomeruliflora Persoon

Despite the common name, *Baccharis* is usually a shrub and only rarely a small tree. The elliptic to obovate leaves are 3–6 cm long, and the small, rayless, sessile flower heads are in clusters scattered along the leafy branches.

A common shrub of fresh or brackish marshes around the entire coast of Florida and on to North Carolina.

S-5A/SES/KTW
(L 179-31-3; R 179-43-2)
ASTERACEAE

CRB

458. Sea Myrtle

Baccharis halimifolia Linnaeus

The numerous flower heads of these dioecious shrubs or small trees are in large, conspicuous, terminal leafy clusters. The elliptic to obovate leaves, 3–6 cm long, are serrate above the middle.

A common plant of old fields, woodland margins, roadsides, and disturbed areas throughout Florida and along the coast to Texas and Massachusetts.

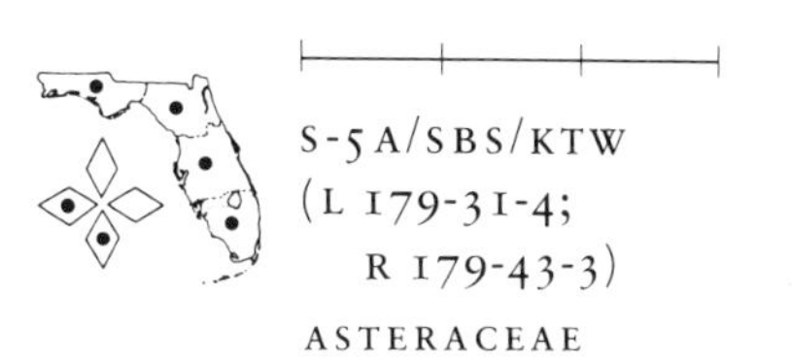

S-5A/SBS/KTW
(L 179-31-4;
R 179-43-3)
ASTERACEAE

WSJ

459. Honeycomb Head

Balduina uniflora Nuttall

The single, usually unbranched, pubescent flowering stem of these perennials can be from 3–10 dm tall. The petiolate linear or elliptic basal leaves are up to 10 cm long. The twelve to twenty yellow rays of each flower head are 2–3 cm long and toothed at the end. (The rays of the related *B. angustifolia*, which gets into south Florida, are only 1–2 cm long.)

Native to the savannas, sandhills, and pineland scrub of northern Florida and along the coast to Louisiana and the Carolinas.

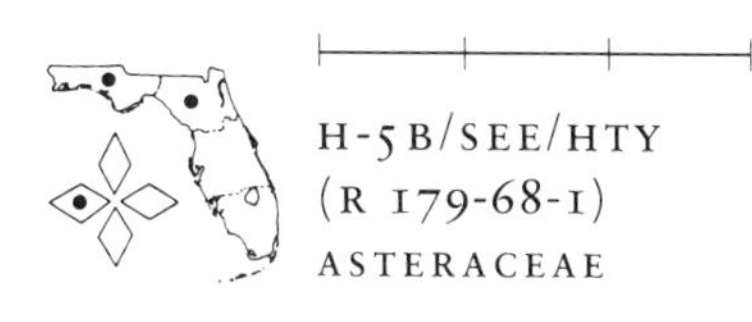

H-5B/SEE/HTY
(R 179-68-1)
ASTERACEAE

BJT

460. Greeneyes

Berlandiera subacaulis (Nutt.) Nuttall

The rough-pubescent, narrowly elliptic basal leaves of this perennial are 10–15 cm long and coarsely toothed or lobed. The common name comes from the greenish yellow central disc flowers of the solitary flower heads. (The related *B. pumila*, of northern Florida, has a maroon disc.)

Greeneyes occurs only in Florida; it is fairly common in pinelands and disturbed areas of south and central Florida, less frequent in north Florida, and known in the Panhandle only from Jefferson County.

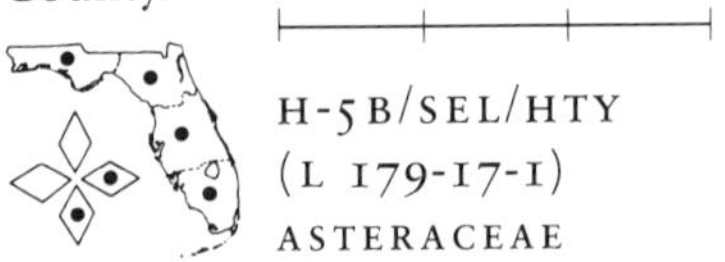

H-5B/SEL/HTY
(L 179-17-1)
ASTERACEAE

BJT

461. Spanish Needles

Bidens pilosa Linnaeus

The 2–6 cm long leaves of these annuals may be entire but are more commonly pinnately divided into narrow elliptic segments. The flower heads normally have five white rays 1–1.5 cm long. (The small rays of *B. bipinnata* are less than 5 mm long.)

A common pantropical weed of roadsides, waste areas, and disturbed soils throughout Florida and on into coastal Louisiana and the Carolinas.

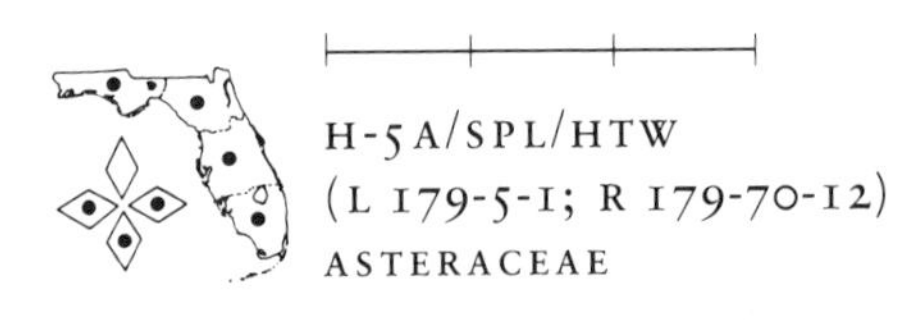

H-5A/SPL/HTW
(L 179-5-1; R 179-70-12)
ASTERACEAE

BJT

462. Bigelowia

Bigelowia nudata (Michx.) de Candolle

The numerous, small, cylindrical flower heads of Bigelowia, each with only two to six tubular flowers, are about 1 cm long and are arranged in open terminal clusters on the 2–8 dm tall flower stalks. The basal leaves are linear and up to 1 dm long.

Bigelowia occurs in moist pinelands and wet meadows in all sections of Florida but varies considerably in frequency; it is also found near the coast, north to the Carolinas and west to Louisiana.

H-5B/LSE/KTY
(L 179-33-1; R 179-50-1)
ASTERACEAE

BJT

463. Sea Daisy

Borrichia frutescens (L.) de Candolle

The yellow central disc of the solitary flower head of these rather coarse perennials is 1–2 cm in diameter, often larger than the yellow, 1-cm-long rays. The elliptic to oblanceolate fleshy leaves are 2–7 cm long.

A common plant of coastal sands and marshes from Virginia to Texas and likewise throughout coastal Florida.

H-5A/SEE/HTY
(L 179-9-1; R 179-64-1)
ASTERACEAE

BJT

464. Deer Tongue

Carphephorus paniculatus (J.F. Gmel.) Herbert

The numerous, small, clustered flower heads, arranged in a branching raceme or panicle on the 1-meter-tall, pubescent flower stalk, are about 1 cm long, and each is composed of three to twelve lavender florets or small tubular flowers. (The similar Vanilla Plant, *C. odoratissimus*, has a fragrant vanilla odor.)

A frequent perennial of sandhills and pine flatwoods in all sections of Florida and north along the coast into North Carolina.

BJT

465. Thistle

Cirsium horridulum Michaux

These stout biennials with horrible spiny leaves may be a meter or more tall, and the large rayless flower heads, which may range from cream or yellow to pink or dark rose, are 5–6 cm or more in diameter.

A frequent native weed of pinelands, pastures, and disturbed areas in all sections of Florida and much of the nonmountainous area of the eastern United States.

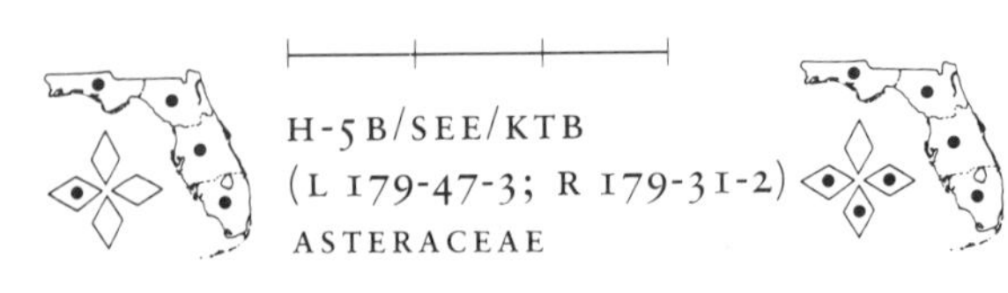

H-5B/SEE/KTB
(L 179-47-3; R 179-31-2)
ASTERACEAE

H-5A/SNL/KTR
(L 179-54-1; R 179-25-3)
ASTERACEAE

CRB

RCS

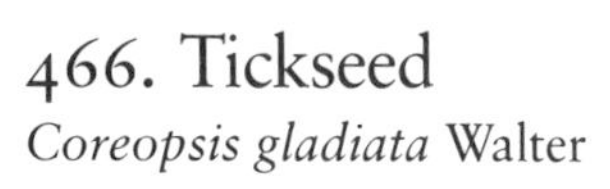

466. Tickseed

Coreopsis gladiata Walter

The small, tubular, maroon or brown disc flowers of this glabrous perennial have only four minute corolla lobes instead of the five usually associated with flowers of the Aster family. Nonetheless, they offer a colorful contrast with the rich yellow, 1.5–2.5 cm long rays, which are notched or toothed at the end.

These tall plants grow in moist to wet open pinelands at scattered localities in all sections of Florida and along the coast to Texas and Virginia.

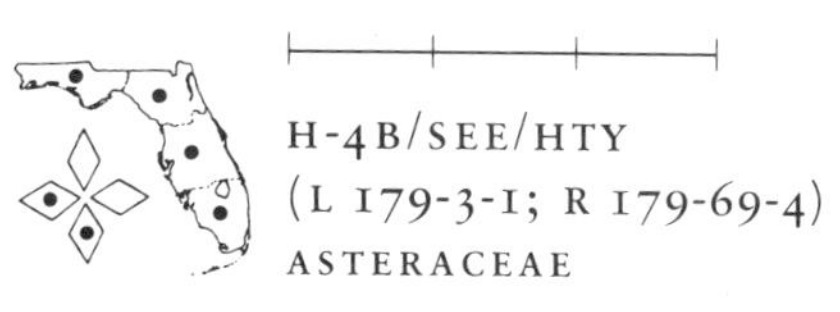

H-4B/SEE/HTY
(L 179-3-1; R 179-69-4)
ASTERACEAE

467. Swamp Coreopsis

Coreopsis nudata Nuttall

The pink rays of these attractive "flowers," which look very much like the cultivated Cosmos to which they are related, are 1.5–3 cm long, and the disc of small yellow tubular flowers (the outer ones just opening in the photograph) is about 1 cm in diameter.

Swamp Coreopsis is a perennial of moist open pinelands and wet ditches of northern Florida and on into Georgia and Louisiana. An especially spectacular display of these and other spring savanna plants may be seen on the Wilma savannas in Liberty County.

H-5B/SLE/HTR
ASTERACEAE

WSJ

468. Elephant Foot

Elephantopus nudatus Gray

The 4–8 dm tall, leafless, hairy scape or flowering stem of this perennial herb rises from a flat basal rosette of large, pubescent leaves 8–25 cm long and 2–6 cm wide. The small heads of tubular flowers have three leaflike bracts just below them. (The leaves of the similar *E. elatus*, common throughout Florida, are more pubescent beneath; the stems of *E. carolinianus* are leafy, the basal leaves absent.)

More frequent in dry deciduous woodlands than in pinelands of central and northern Florida and much of the Southeast.

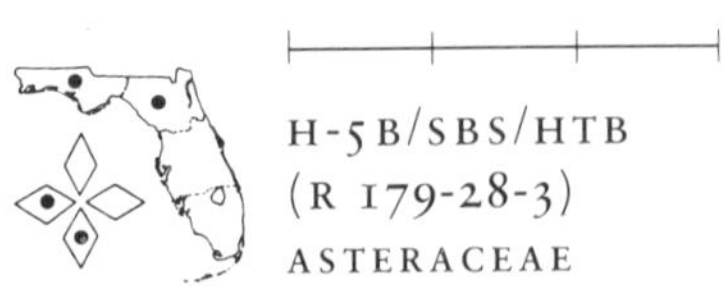

H-5B/SBS/HTB
(R 179-28-3)
ASTERACEAE

BJT

469. Tasselflower

Emilia fosbergii D.H. Nicolson

The urceolate, or urn-shaped, flower heads of these introduced annuals are about 2 cm long, and the plants may be 2–5 dm or more tall. (Two other very similar old-world species also naturalized in our area, *E. sonchifolia* and *E. coccinea*, are separated by small differences in the involucre, the green series of bracts around the base of the flower head.)

These pantropical weeds are established in vacant lots, roadsides, lawns, and waste areas of south and central Florida.

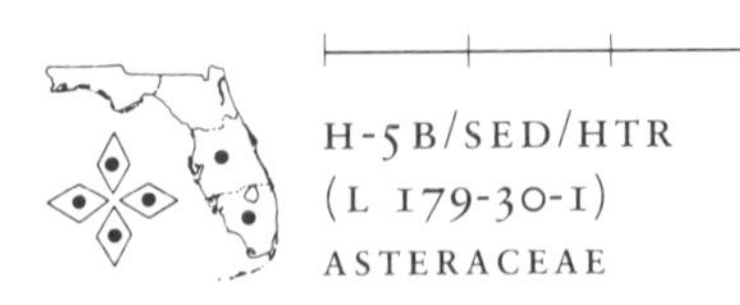

H-5B/SED/HTR
(L 179-30-1)
ASTERACEAE

BJT

470. Southern Fleabane
Erigeron quercifolius Lamarck

Each of the numerous small flower heads of this Fleabane have 100 to 200 fine, linear, blue lavender rays about 5 mm long. These freely branched, pubescent herbs are 2–7 dm tall and have obovate, shallowly lobed basal leaves and clasping stem leaves. (The heads of the similar *E. vernus* have less than fifty rays; the stem leaves of Daisy Fleabane, *E. strigosus*, are sessile but not clasping.)

Common weeds of old fields and disturbed sites in all sections of Florida, and on the coastal plain to North Carolina and Texas.

H-5A/SOD/HTB
(L 179-39-2; R 179-44-3)
ASTERACEAE

WSJ

471. Dog Fennell
Eupatorium capillifolium (Lam.) Small

The numerous small branches, the small linear or finely divided opposite or often fascicled leaves, and the multitude of relatively minute flower heads combine to give these rank 1–2 meter tall herbs a distinctive feathery appearance. (The upper leaves of the very similar *E. compositifolium* are 1 mm or more wide.)

These weedy perennial herbs are common throughout Florida and, primarily, the coastal areas of other southeastern states.

H-5O/SLE/HTW
(L 179-49-4; R 179-34-5)
ASTERACEAE

BJT

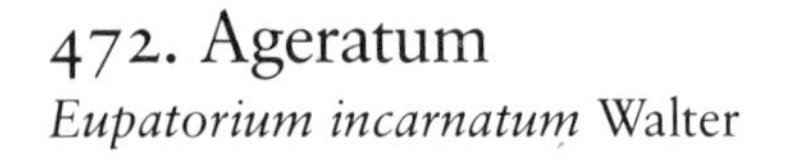

472. Ageratum

Eupatorium incarnatum Walter

Although there are only eighteen to thirty small pink tubular flowers in each flower head, the large number of these heads, terminal on the numerous branches of these 3–10 dm tall perennials, make these plants quite showy. The opposite, petiolate, triangular leaves are 4–7 cm long. (The very similar *E. coelestinum* from which a number of horticultural varieties have been developed, has thirty-five to seventy flowers per head.)

Infrequent along ditch banks and woodland margins of only north Florida, but throughout much of the rest of the Southeast.

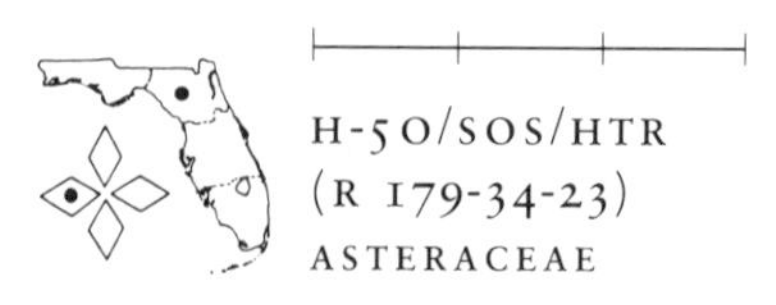

H-50/SOS/HTR
(R 179-34-23)
ASTERACEAE

BJT

473. Late Thoroughwort

Eupatorium serotinum Michaux

The opposite ovate, petiolate leaves of this *Eupatorium* can be 6–12 or more cm long and have small resinous dots beneath. (The related *E. aromaticum*, also with white flower heads, is fragrant when crushed.)

These perennial herbs are frequent in open deciduous woods, clearings, and waste areas of Florida and much of the eastern United States.

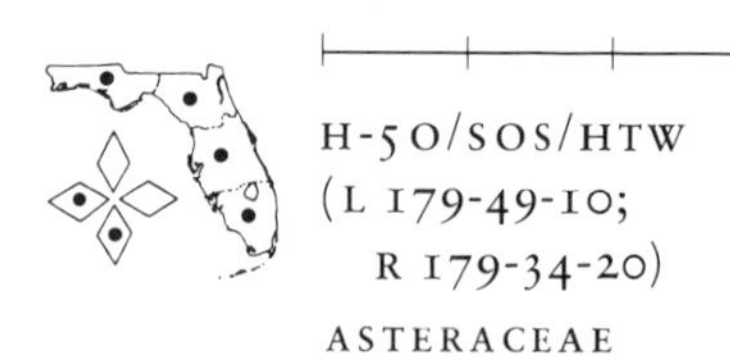

H-50/SOS/HTW
(L 179-49-10;
R 179-34-20)
ASTERACEAE

BJT

474. Flat-topped Goldenrod

Euthamia minor (Michx.) Greene

The common name adequately describes these freely branched, 3–10 dm tall, rhizomatous perennials. The small flower heads are composed of ten to sixteen rather minute ray flowers and five to seven equally small disc flowers. The linear, gland-dotted leaves are 1–2 mm wide. (The leaves of the similar *E. tenuifolia*, rare in our area, are 2–5 mm wide.)

Frequent throughout Florida in open sandy meadows, pinelands, and disturbed sites, and on the coastal plain to Maryland and Louisiana.

H-5A/SLE/KTY
(L 179-34-2;
R 179-49-38)
ASTERACEAE

BJT

475. Yellowtop

Flaveria linearis Lagasca

The opposite leaves of these 3–10 dm tall, often semishrubby, perennial herbs are linear and 4–15 cm long. The numerous but small flower heads may each consist of as few as one ray flower and one disc flower. (The heads of *F. floridana*, also in our area, have ten to fifteen florets.)

Plants of coastal hammocks, dunes, pinelands, and disturbed areas of central and south Florida, the West Indies, and Mexico.

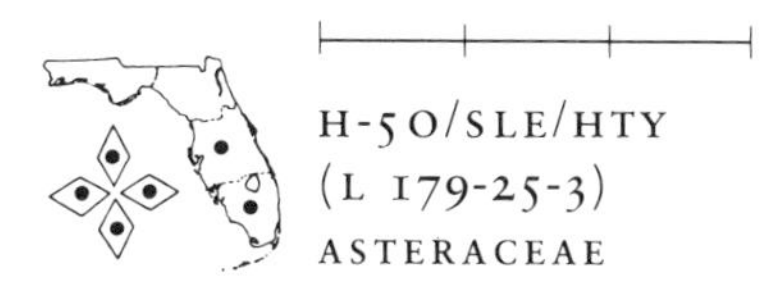

H-5O/SLE/HTY
(L 179-25-3)
ASTERACEAE

BCB

476. Gaillardia

Gaillardia pulchella Fougeroux

The dark central disc of the attractive flower heads of these somewhat coarse annuals is 1–2 cm in diameter, and the six to fifteen usually bicolored rays are 1–2 cm long. The leaves may be entire or pinnately cut. (The similar *G. aestivalis*, which ranges south to Highlands County, has fine rather than coarse pubescence.)

A frequent plant of sandy open sites, especially near the coast, throughout Florida and much of the Southeast.

H-5A/SEE/HTR
(L 179-22-1; R 179-74-1)
ASTERACEAE

BJT

477. Garberia

Garberia fruticosa (Nutt.) Gray

The elliptic to obovate, entire leaves of this aromatic, semi-evergreen shrub, which may be 1–2.5 meters tall, are 1.5–3.5 cm long and have small glandular dots.

A relatively frequent endemic component of the pine-oak scrub vegetation of northern and central Florida.

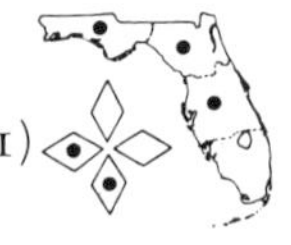

S-5A/SEE/KTB
ASTERACEAE

CRB

478. Rabbit Tobacco

Gnaphalium obtusifolium Linnaeus

The fully branched, 3–10 dm tall stems of these biennials have a tawny to silver white pubescence, and the narrow, linear aromatic leaves, mostly 3–6 cm long, are white-tomentose beneath. Dried plants of Rabbit Tobacco are sometimes used in dried flower arrangements, thus the name Sweet Everlasting. (The flower heads of the related, and less profusely branched, *G. purpureum*, are in spikes.)

A common weed of old fields, pastures, roadsides, and waste places in all sections of Florida and much of the Southeast.

H-5A/SLE/KTW
(L 179-43-1; R 179-40-1)
ASTERACEAE

WSJ

479. Bitterweed

Helenium amarum (Raf.) H. Rock

The five to ten rays of each flower head are 5–12 mm long, about the diameter of the yellow central disc. The leaves of these annuals are filiform or have filiform segments. Even in pastures grazed to the bare ground, plants of Bitterweed will not be touched. (The leaves of *H. flexuosum*, also in our area, are lanceolate and the stems winged.)

A common weed of pastures, roadsides, and waste places throughout Florida and much of the Southeast.

H-5A/SLE/HTY
(L 179-26-3; R 179-75-6)
ASTERACEAE

BCB

480. Beach Sunflower

Helianthus debilis Nuttall

The ten to twenty rays of each flower head of these variable annuals are 1–3 cm long, and the maroon central disc is 1–2.5 cm in diameter. The ovate to cordate leaves are often scabrous or rough to the touch. (The common cultivated Sunflower, *H. annuus*, native to the Midwest, is an occasional escape in our area.)

Although these are primarily coastal dune plants, they also occur inland on disturbed soils and are found throughout Florida and along the coast to Texas and the Carolinas.

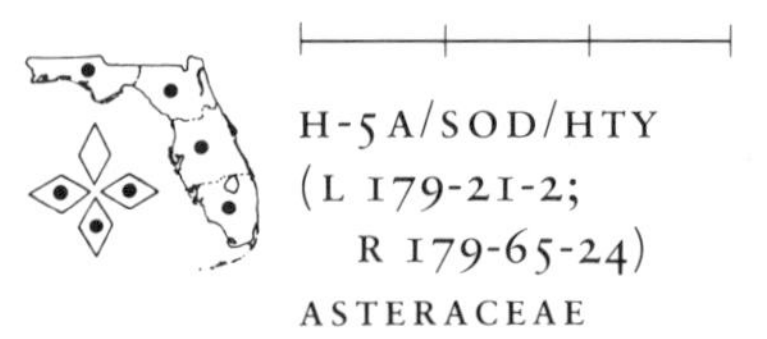

H-5A/SOD/HTY
(L 179-21-2;
R 179-65-24)
ASTERACEAE

CRB

481. Rayless Sunflower

Helianthus radula (Pursh) Torrey & Gray

The large, flat, obovate basal leaves, 4–10 cm long, and the tall, hairy, nearly leafless flower stalk with a solitary rayless flower head make this one of our most atypical and distinctive Sunflowers.

These native perennials often form large colonies along roadsides and in moist pinelands in all parts of Florida and along the coast to Virginia and Louisiana.

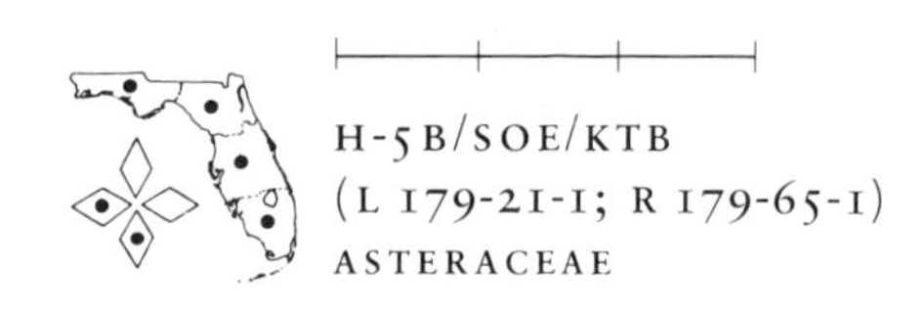

H-5B/SOE/KTB
(L 179-21-1; R 179-65-1)
ASTERACEAE

BJT

482. Cat's Ear

Hypochoeris radicata Linnaeus

The Dandelion-like flowers of this introduced old-world perennial are about 2 cm in diameter, and the involucre, or green cuplike base of the flower head, is 1–1.5 cm long.

This weed is well established in lawns, pastures, and waste places over much of the eastern United States and south into northern Florida.

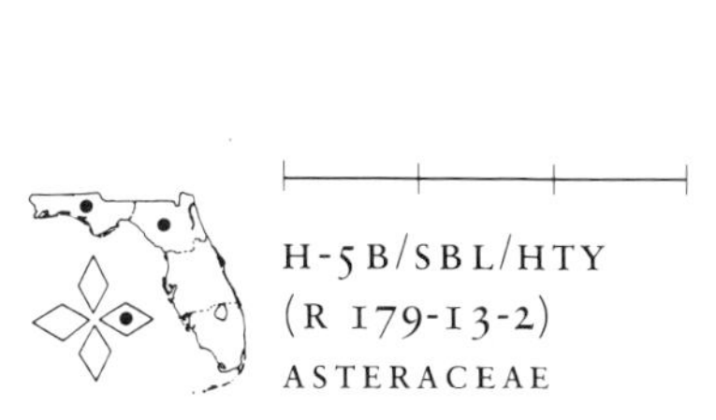

H-5B/SBL/HTY
(R 179-13-2)
ASTERACEAE

BJT

483. Blazing Star

Liatris tenuifolia Nuttall

The long spikes of colorful flower heads of *Liatris*, which bloom from the top down, may be from 1–5 dm or more long and the entire flower stalk 1–2 meters tall. One or more of the dozen or so rather similar species of these lavender-flowered perennials occcur in various areas of Florida.

Frequent in dry pinelands, sandhills, roadsides, and clearings throughout our state and on the coastal plain to South Carolina and Louisiana.

H-5A/SLE/KTB
(L 179-50-5; R 179-30-9)
ASTERACEAE

BJT

484. Roserush

Lygodesmia aphylla (Nutt.) de Candolle

The rose lavender floral rays and the leafless, rushlike stem of these perennials give it the common name. The flower head is 2–3 cm in diameter; the few linear basal leaves, 1–3 dm long, are often absent by flowering time.

A relatively frequent but local plant of dry pinelands and sand scrub of Florida and southern Georgia.

H-5B/SLE/HTR
(L 179-55-1)
ASTERACEAE

DNP

485. Barbara's Buttons

Marshallia tenuifolia Rafinesque

The rose-colored flowering heads on slender, usually branched stems 3–8 dm tall are composed of only tubular disc flowers and are 2–4 cm in diameter. The linear stem leaves are almost erect.

These somewhat infrequent perennials are native to savannas and clearings of central and northern Florida, southern Georgia, and west to Texas.

H-5A/SLE/KTR
ASTERACEAE

WSJ

486. Climbing Hempweed

Mikania scandens (L.) Willdenow

The small clusters of somewhat drab rayless flower heads are numerous but not showy. The cordate, acuminate, or sharply pointed, leaves are 3–12 cm long. This twining perennial vine often forms thick mats over other low vegetation. (The very similar *M. cordifolia* also occurs in south and central Florida.)

Common in wet thickets, fencerows, and clearings throughout Florida and the southeast coastal plain from Virginia to Mississippi.

V-5O/SCE/KTR
(L 179-46-3; R 179-35-1)
ASTERACEAE

BJT

487. Palafoxia

Palafoxia feayi Gray

The white to very pale lavender flower heads, composed only of seventeen to thirty tubular disc flowers, are 2–3 cm in diameter, and the numerous elliptic stem leaves are 2–6 cm long. (The lower leaves of the more widespread *P. integrifolia* are linear, rather than ovate-lanceolate as in this species.)

These perennials are another endemic to the dry pinelands and scrub of central and south Florida.

H-5A/SEE/KTW
(L 179-23-1)
ASTERACEAE

DNP

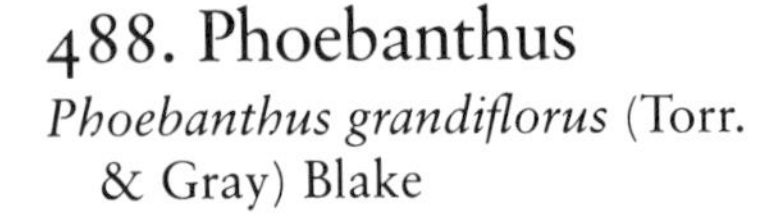

488. Phoebanthus

Phoebanthus grandiflorus (Torr. & Gray) Blake

The flower heads of these rhizomatous, perennial herbs have ten to fifteen rays, 2–4 cm long, around a yellow disc 1.5–2.5 cm in diameter. The sparsely branched flower stalks are 5–10 dm tall. (The leaves of the similar but rare *P. tenuifolius*, endemic to a small area near the Appalachicola River, are less than 3 mm wide.)

A relatively frequent sandhill endemic in central Florida that is rare in south Florida.

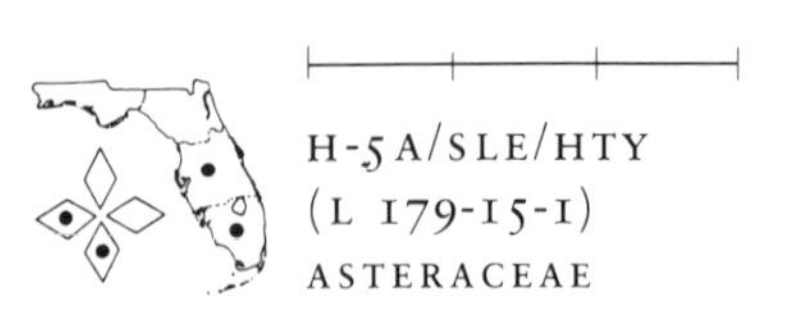

H-5A/SLE/HTY
(L 179-15-1)
ASTERACEAE

BJT

489. Golden Aster

Pityopsis graminifolia (Michx.) Nuttall

These branched, silvery pubescent perennials are 4–8 dm tall, have linear leaves 1–3 dm long, and numerous golden yellow flower heads, 1.5–3.5 cm in diameter, with four to ten rays. (The wider leaves of the related and similar *Heterotheca subaxillaris* are ovate.)

Common in dry pinelands, old fields, and on sandhills throughout Florida, much of the eastern United States, and islands of the West Indies.

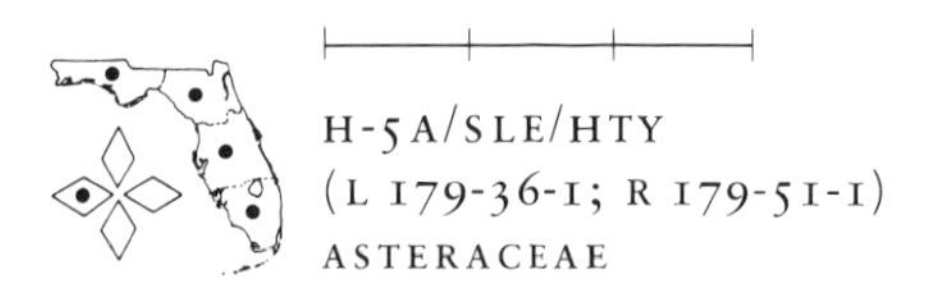

H-5A/SLE/HTY
(L 179-36-1; R 179-51-1)
ASTERACEAE

BJT

490. Camphor Weed

Pluchea odorata (L.) Cassini

The ovate to lanceolate, petiolate leaves of this robust annual are 5–12 cm long and more or less succulent. The pale rose rayless flower heads are in rather compact terminal clusters. (The leaves of both the rank-smelling *P. foetida* and the dark-pink-flowered *P. rosea* are sessile; the Mexican *P. symphytifolia*, introduced into south Florida, is a shrub.)

Another name, Salt Marsh Fleabane, is appropriate for these plants that are found in salt marshes along the coast from New England to Mexico and the islands of the West Indies.

H-5A/SNS/KTR
(R 179-36-4)
ASTERACEAE

WSJ

491. Blackroot

Pterocaulon pycnostachyum (Michx.) Ellis

The slender green wings on the silvery pubescent stem and the compact spike of rayless flower heads are positive identification of this colorful perennial, which is usually 3–5 dm tall.

Frequent but often solitary or in sparse populations in the dry open pinelands and sandhills of Florida and on the coastal plain to the Carolinas and Louisiana.

H-5A/SLE/KTW
(L 179-42-1; R 179-41-1)
ASTERACEAE

CRB

492. Black-eyed Susan

Rudbeckia hirta Linnaeus

The dark brown disc flowers contrast sharply with the rich yellow ray flowers, or "petals," of these often cultivated, scabrous annual to perennial herbs. (The disc of *R. laciniata*, of northern Florida, is greenish yellow.)

These colorful weeds occur sporadically in old fields, along roadsides, and in waste areas throughout Florida and much of the eastern United States.

H-5A/SEE/HTY
(L 179-20-2; R 179-61-6)
ASTERACEAE

BJT

493. Goldenrod

Solidago fistulosa P. Miller

These perennial herbs with leafy, pubescent stems may be as much as 2 meters tall with the numerous small flower heads, composed of relatively minute ray and disc flowers, in spikes on slender upper branches. (Several other of our large Goldenrods, including *S. gigantea*, are similar in general appearance but are separated on small floral differences.)

This species is frequent in dry pinelands throughout Florida and along the coast to New Jersey and Louisiana.

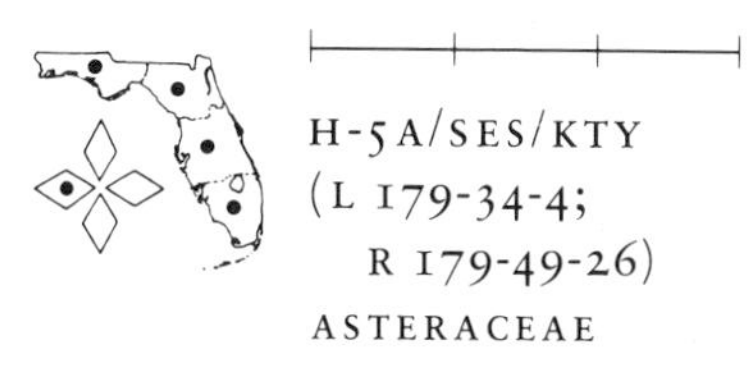

H-5A/SES/KTY
(L 179-34-4;
R 179-49-26)
ASTERACEAE

BJT

494. Slender Goldenrod

Solidago stricta Aiton

These generally unbranched perennial herbs may be from 3–20 dm tall and have thick oblanceolate basal leaves, 5–25 cm long, and greatly reduced cauline, or stem, leaves. The flower heads may have three to seven ray flowers and eight to twelve disc flowers. (The related, but branched, *S. sempervirens* also occurs in salt marshes where the two species have been reported to hybridize.)

Plants of moist pinelands and coastal salt marshes from New Jersey to Mexico and the islands of the West Indies.

H-5A/SBE/KTY
(L 179-34-1; R 179-49-13)
ASTERACEAE

BJT

495. Sow Thistle

Sonchus asper (L.) Hill

The weakly spiny leaves of these rather soft fleshy annuals are auriculate or lobed at the base and clasp the smooth stem, which may be up to 2 meters tall. The flower heads, about 2 cm in diameter, are composed of all ligulate, or raylike, flowers. (The similar *S. oleraceus*, also introduced from Europe, is less spiny.)

A cosmopolitan introduced weed of roadsides, vacant lots, and waste areas throughout Florida.

H-5A/SLS/KTY
(L 179-60-2; R 179-8-2)
ASTERACEAE

WSJ

496. Stokesia

Stokesia laevis (Hill) Greene

The usually solitary showy blue or lavender flower heads of Stokes' Aster, as these perennials are also called, are 4–6 cm in diameter. The thick, glandular-punctate, elliptic basal leaves are 10–30 cm long. Several horticultural varieties of this native species have been developed.

Occasional to rare in moist coastal plain pinelands of northern Florida into Georgia and Louisiana.

H-5A/SEE/HTB
(R 179-23-1)
ASTERACEAE

WSJ

DNP

497. Wild Marigold

Tagetes minuta Linnaeus

These glabrous, aromatic annuals with small, rather cylindrical flower heads, about 1.5 cm long, have pinnately compound leaves with linear, serrate segments. (The related cultivated and more showy "French" Marigold, *T. patula*, and *T. erecta*, the "African" Marigold, both native of Mexico, occasionally escape in our area.)

Wild Marigold, an introduced weed from South America, occurs in old fields and waste places of northern Florida and coastal areas of Alabama, Georgia, and South Carolina.

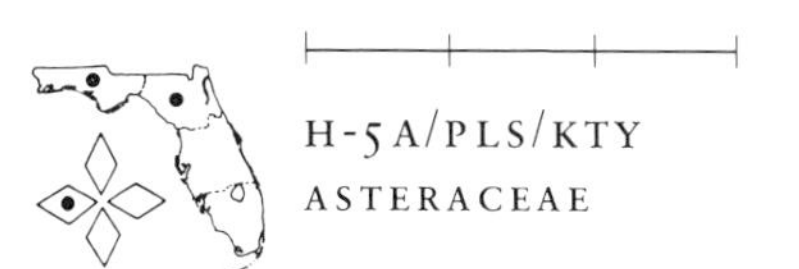

H-5A/PLS/KTY
ASTERACEAE

498. Squarehead

Tetragonotheca helianthoides Linnaeus

The opposite, lanceolate to elliptic, serrate leaves of this caespitose, or clump-forming, perennial herb are 10–16 cm long. The four involucral bracts that enclose the square buds give the plant its common name.

A plant of sandy inner coastal plain woods found in northern Florida (and possibly a few adjacent counties of central Florida) and into Mississippi and the Carolinas.

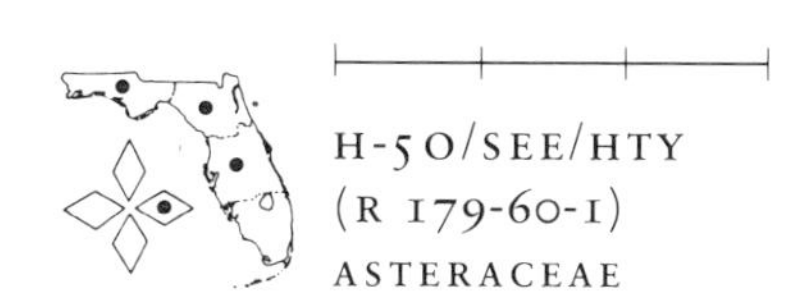

H-5O/SEE/HTY
(R 179-60-1)
ASTERACEAE

BJT

499. Ironweed

Vernonia angustifolia Michaux

The numerous linear leaves of this *Vernonia* are 5–10 cm long but only 2–6 mm wide. The colorful flower heads are 2–3 cm in diameter. (The leaves of Giant Ironweed, *V. gigantea*, which grows in wet meadows, are 3–6 cm wide.)

These native perennials are frequent in dry pinelands and old fields of central and northern Florida and on the coastal plain to Mississippi and the Carolinas.

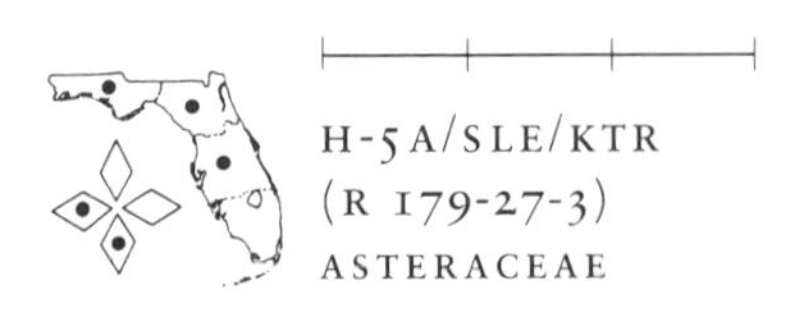

H-5A/SLE/KTR
(R 179-27-3)
ASTERACEAE

BJT

500. Creeping Oxeye

Wedelia trilobata (L.) A.S. Hitchcock

The somewhat fleshy opposite leaves of these low decumbent or prostrate perennial herbs are 5–10 cm long, and the usually solitary flower heads are 3–4 cm in diameter.

Native, or perhaps naturalized from the West Indies, in south and central Florida where it occasionally occurs in open pinelands and disturbed areas.

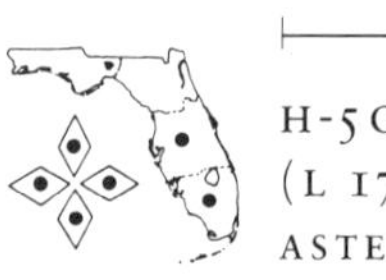

H-5O/SES/HTY
(L 179-6-1)
ASTERACEAE

Glossary

Actinomorphic. *Regular, radially symmetrical; descriptive of a flower, or of a set of flower parts, that can be cut through the center into equal and similar parts along two or more planes.*
Acuminate. *Long-tapering to a pointed apex.*
Angiosperm. *One of the flowering plants (ovules enclosed in an ovulary, or ovary).*
Annual. *Life cycle completed in one year or less.*
Anther. *The fertile part of the stamen; the part that produces the pollen.*
Auriculate. *An ear-shaped appendage or lobe.*
Axil, leaf. *The angle formed between the stem and the upper surface of the leaf.*
Axillary. *In an axil.*
Banner. *The upper petal, or standard, of a papilionaceous flower, as in the legumes (#315).*
Biennial. *A plant that completes its life cycle in two years and then dies; normally remains as a vegetative rosette the first year, flowering the second year.*
Biblabiate. *Two-lipped, as some corollas that are thus zygomorphic.*
Blade. *The expanded or flattened part of the leaf.*
Bract. *A reduced leaf, particularly one subtending a flower, or inflorescence, as the involucral bracts in the Aster family.*
Bud. *An unopened flower or cluster of embryonic leaves, often enclosed by scales.*
Bulb. *A short underground stem surrounded by fleshy leaves or scales.*
Bulbil. *A small bulb-like body, especially those borne on a stem or in an inflorescence, as in* Allium *(#22).*
Buttress. *Additional, often flattened, supporting tissue at the base of the trunk, as in Cypress (#16).*
Campanulate. *Bell-shaped, or funnel-shaped, as some corollas.*
Capsule. *A dry, dehiscent fruit derived from two or more united carpels.*
Catkin. *A scaly bracted, usually flexuous spike or spike-like inflorescence, often of unisexual flowers, as in* Alnus *(#274).*
Cauline. *Occurring along a stem, as opposed to basal.*
Cespitose. *Tufted in clumps.*

Clasping. *The base partly or wholly surrounding another structure, as some leaves clasp stems.*

Cleft. *Cut ¼ to ½ the distance from the margin to midrib, or apex to base, or generally, any deep lobe or cut.*

Cleistogamous. *A type of self-pollinated flower that does not open.*

Colonial. *A group of plants with a clonal relationship where all plants are from one rootstock, rhizome, stolon, or root system.*

Column. *The united style and filaments in the Orchidaceae.*

Coriaceous. *With a leathery texture.*

Corm. *A bulb-like underground structure in which the fleshy portion is predominantly stem tissue and is covered by membranous scales.*

Corolla. *The part of the flower made up of petals.*

Corymb. *Short, broad, more or less flat-topped, indeterminate inflorescence, the outer flowers opening first.*

Culm. *The flowering stems of grasses and sedges.*

Cyathium. *The specialized involucrate inflorescence of* Poinsettia *(#197).*

Cyme. *An inflorescence that has a series of more or less equal branches that bear the flowers.*

Deciduous. *Trees or shrubs that shed all their leaves each year; not evergreen.*

Decumbent. *Reclining or lying on the ground, but with the end ascending.*

Dentate. *Toothed, the sharp or coarse teeth perpendicular to the margin.*

Dicotyledons. *Plants in one of the two subgroups of the angiosperms; characterized by two cotyledons, or embryo leaves.*

Dioecious. *Having the male and female reproductive organs on separate plants.*

Disc. *Enlarged outgrowth of receptacle; in Asteraceae, the tubular flowers.*

Divided. *Cut ¾ or more the distance from the leaf margin to midrib, or petal apex to base, or generally, any deep cut.*

Drupe. *A fleshy, usually 1-seeded indehiscent fruit with seed enclosed in a stony endocarp.*

Emarginate. *Having a shallow notch at the end.*

Endemic. *Restricted to a relatively small area or region.*

Entire. *A margin without teeth, lobes, or divisions.*

Ephemeral. *Lasting only a short time.*

Epiphytic. *A plant growing upon another plant, but not as a parasite.*

Evergreen. *Green all year, not shedding all of its leaves at one time.*

Farinose. *Covered with a whitish mealy powder.*

Fertile. *A flower, or flower part, bearing functional reproductive structures.*

Fiddlehead. *The curled young fronds, or leaves, of a fern (#7).*
Filiform. *Thread-like, slender and usually round in cross section.*
Flora. *A collective term to refer to all of the plants of an area; a book dealing with the plants of an area.*
Flower. *The characteristic reproductive structure of angiosperms; typically with 4 sets of parts (calyx, corolla, stamens, and pistil); see Figure 3. An incomplete flower has one or more sets of parts missing.*
Follicle. *A dry fruit, from a single ovary, that splits open along a single line.*
Frond. *The leaf of a fern, often appearing compound or decompound.*
Fruit. *A matured ovulary with or without accessory structures.*
Glabrous. *Without trichomes or hairs.*
Glandular. *Having or bearing secreting organs, glands or trichomes.*
Glaucous. *Covered with a thin, whitish, waxy "bloom" that can be wiped off.*
Herb. *A plant with no persistent woody stem above ground.*
Herbceous. *Plant parts with little or no hard, woody (secondary) tissue.*
Hirsute. *With rather rough or corase trichomes, or hairs.*
Hypanthium. *A structure formed by the fusion of the sepals, petals, and stamens, as in the Rosaceae.*
Indeterminate. *Not of limited growth or size; not determinate.*
Inflorescence. *The flowering section of a plant.*
Involucre. *A whorl or collection of bracts surrounding or subtending a flower cluster or a single flower.*
Knees. *A rounded or spurlike process rising from the roots of certain swamp-growing trees, as in cypress.*
Leaf. *The flattened, usually green, vegetative organ consisting of a distal blade (the flattened part) and a stalk or petiole.*
Leaf, compound. *A leaf in which the blade is subdivided into two or more leaflets; compound leaves may be pinnate, bipinnate, or palmate.*
Leaf, simple. *A leaf with only one blade, not compound.*
Leaflet. *A single unit or division of a compound leaf which will ultimately separate from the leaf axis by an abscission layer.*
Legume. *A simple, dry, dehiscent fruit splitting along two sutures; characteristic of the Fabaceae, or bean family (#283).*
Linear. *Long and narrow with essentially parallel margins, as the blades of most grasses.*
Lip. *The upper or lower portion of a two-lipped, or zygomorphic, flower.*
Lobed. *Cut from 1/8 to 1/4 the distance from the margin to midrib, or apex to base, or more generally, any cut resulting in rounded segments.*
Membranous. *Thin, membrane-like.*

Monocotyledon. *Plants in one of the two subgroups of the angiosperms; characterized by one cotyledon in the embryo.*

Monoecious. *Having both kinds of incomplete (unisexual) flowers borne on a single plant as in* Cattail *(#78).*

Node. *The point on a stem at which one or more leaves are produced.*

Oblique. *Slanting; unequal-sided.*

Obovate. *Inversely ovate.*

Ovary. *The basal part of the pistil that bears the ovules which, upon fertilization, produce the seeds.*

Panicle. *An indeterminate branching raceme; an inflorescence in which the branches of the primary axis are racemose and the flowers pedicellate.*

Parallel venation. *Leaf venation in which the major veins (vascular bundles) are parallel with one another; relatively characteristic of monocotyledons.*

Parasite. *A plant that gets its food from another living organism.*

Parted. *Cut from 1/2 to 3/4 the distance from the margin to midrib or apex to base, or more generally, any moderately deep cut.*

Pectinate. *Divided into long, linear, equal, lateral divisions; comb-like.*

Pedicel. *The stalk of an individual flower in an inflorescence.*

Peduncle. *The stalk of a flower cluster or of a single flower if the inflorescence consists of a single flower.*

Peltate. *With the petiole joining the blade near the center rather than at the margin, as in* Hydrocotyle *(#418).*

Pendulous. *Drooping, hanging downward.*

Perennial. *Plant of three or more years duration.*

Perfoliate. *A sessile leaf or bract whose base completely surrounds the stem, the latter seemingly passing through the leaf (see #424).*

Perianth. *The calyx and corolla of a flower.*

Petal. *One of the leaf-like appendages making up the corolla.*

Petiole. *The stem-like part of the leaf.*

Pistil. *The female reproductive parts of a flower; the stigma, style, and ovary collectively.*

Pistillate. *A flower with one or more pistils but not stamens; a female flower, flower part, or plant.*

Plicate. *Folded, as a paper fan.*

Pollination. *The actual transfer of pollen grains from the anther of the stamen to the stigma of the pistil.*

Prop Root. *Any root which serves as a prop or support to the plant, as in maize or the mangrove.*

Prostrate. *A general term for lying flat on the ground.*

Pseudobulb. *The solid, bulbous enlargement of the stem found in many epiphytic orchids.*

Pubescence. *A general term for hairs or trichomes.*

Pubescent. *Covered with short, soft trichomes.*
Punctate. *With translucent or colored dots, depressions, or pits.*
Rachis. *A stem bearing flowers or leaflets.*
Ray. *A single branch of an umbel; the strap-shaped ligulate, or "petal", flowers in the inflorescence of the Asteraceae.*
Reflexed. *Abruptly recurved or bent downward or backward.*
Resinous. *With the appearance of resin; glandular dotted.*
Rhizome. *An underground stem, usually horizontally oriented and sometimes specialized for food storage.*
Rosette. *An arrangement of leaves radicating from a crown or center and usually at or close to the earth, as in* Berlandiera *(#460).*
Rotate. *Flat, and round in outline, without an elongate tube.*
Sagittate. *Like an arrowhead in form; triangular, with the basal lobes pointing downward or inward toward the petiole.*
Saprophyte. *A plant that gets its food from dead organic material.*
Scabrous. *Rough, feeling rough or gritty to the touch.*
Scape. *A leafless or naked flowering stem.*
Scapose. *Producing a scape.*
Sepals. *The outermost, sterile, leaf-like parts of a complete flower.*
Serrate. *With sharp teeth pointing forward.*
Sessile. *Without petiole or pedicel.*
Sorus. *The cluster of sporangia, or spore cases, of ferns.*
Spadix. *A spike with a fleshy axis in which the flowers are imbedded, as in Araceae (#73).*
Spathe. *A large bract enclosing or surrounding an inflorescence.*
Spatulate. *Oblong, but attenuated to base, spoon-like.*
Sporangium. *A spore case.*
Spore. *A unicellular, asexual reproductive unit of ferns and lower plants.*
Spur. *A tubular or sac-like projection from a petal or sepal; also, a very short branch with compact leaf arrangement.*
Stamen. *The pollen-bearing structure in a flower.*
Staminate. *A flower with stamens but no pistil; a male flower or plant.*
Stigma. *The pollen-receptive, terminal part of the pistil.*
Stipule. *The basal, paired, leaf-like appendages of a petiole, sometimes fused.*
Style. *The elongated, sterile portion of the pistil between the stigma and the ovary.*
Subtending. *Below or beneath, as the bracts subtending an inflorescence.*
Succulent. *Juicy, fleshy.*
Suture. *A seam, or a line of opening.*
Tendril. *A slender twining appendage or axis that enables plants to climb.*

Terminal. *At the tip, apical, or distal.*
Ternate. *In three; three-parted or divided, as some leaves.*
Tomentose. *With tomentum; densely woolly or pubescent; with matted, soft, wool-like trichomes.*
Trichome. *A plant hair; trichomes may be simple, stellate, or glandular.*
Trifoliate. *Having three leaves or leaflets (trifoliolate).*
Tuber. *A fleshy; enlarged portion of a stem, rhizome or stolon with only vestigial scales; true tubers are found in the Solanaceae.*
Tubular. *Having the form or consisting of a tube or tubes.*
Unifoliolate. *Compound with but a single leaflet as in the orange leaf; distinguished by the basal joint.*
Urceolate. *Urn-shaped, as the corollas of some Ericaceae (see #153).*
Verticillate. *Arranged in a whorl.*
Villous. *Provided with long, soft, shaggy trichomes.*
Whorl. *Three or more leaves or flowers at one node; in a circle.*
Whorled leaf arrangement. *Three or more leaves attached to a stem at a single node.*
Wing. *A thin membranous extension; the lateral petals in Fabaceae and Polygalaceae.*
Zygomorphic. *A bilaterally symmetrical flower which is divisible into equal halves in one plane only, usually along a vertical line; not actinomorphic or radially symmetrical.*

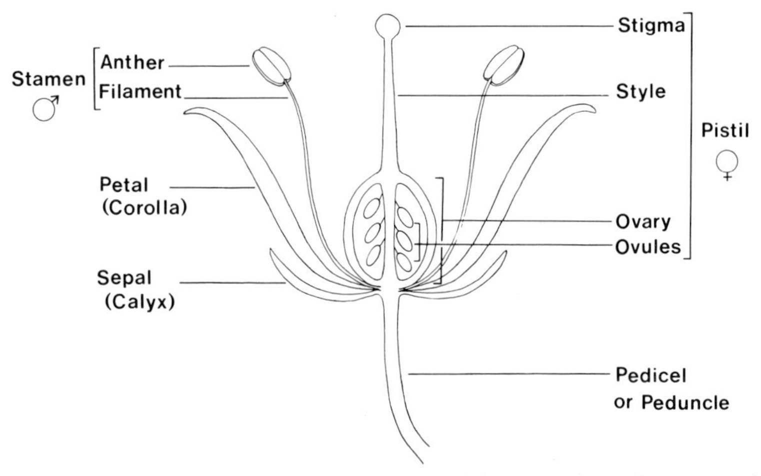

FIGURE 3. Diagramatic section through a typical flower to show the parts and their relative positions.

Identification Keys

Plants are often identified by a "key", or a series of paired statements about the flowers, leaves, or other aspects of the plants, that are answered either "yes" or "no". The series ultimately ends with the name of a plant, or group of plants, which have all of the characteristics indicated by the sequence of "yes" answers. In its basic structure such a paired, or dichotomous key, is the same as the binary system used in computers.

If all of the plants of an area are included in the key, and if the correct information (or interpretation!) is used for each choice, almost any plant in the area covered by the key can be quickly and accurately identified with only occasional reference to a pertinent glossary of the more common botanical terms.

Although the following keys are specifically for the 500 plants illustrated in this book, they may also provide means for the general identification of one half or more of Florida's more obvious wild flowers and roadside plants. However, it is not always obvious whether or not a particular plant is included in the picture section until you try to "run the plant through the key" and then check the plant in question against the appropriate picture or pictures in the book.

For example, if you find a pink or red waterlily you would start with Key R-2: plants with pink or red flowers that are aquatic. This is a short key; pink waterlilies are not illustrated in the book and are thus not included in the key, so you know at once that the plant you have cannot be identified completely just on the basis of pictures or keys presented here. On the other hand, the rather frequent yellow-flowered roadside weed, *Descurania pinnata*, or Tansy Mustard, which is *not* pictured in the book, is similar enough in its key characters to the closely related Wild Radish (#184, in the Brassicaceae, or Mustard family) that it will easily key out to that name. Only a comparison of the actual plant with the picture will show that the two plants are not the same and that further checking in a more inclusive treatment of the region's flora will be necessary for an accurate identification.

A second point to remember when you key out a plant is that a certain amount of variation is normal, and universal, in all biological material: a species with normally pink or blue flowers may have a mutant form with white flowers, or flower color may change with age; a species that has spiny stems, such as Sky-flower, or *Hydrolea* (#428) in Key

	Flowers Absent	Green	White, Cream	Yellow, Orange	Pink, Red	Blue, Purple, Brown
Tree, Shrub, Vine	A	G-1	W-1	Y-1	R-1	B-1
Aquatic, Epiphyte, Parasite	A	G-2	W-2	Y-2	R-2	B-2
Terrestrial Herb	A	G-3	W-3	Y-3	R-3	B-3

B-3, may occasionally have stems without spines. In many such cases the plant will be keyed out twice in order to compensate for the variation.

And, finally, a caution that many of the most workable keys are based on easily observed vegetative as well a floral characters and thus will tend to group those plants with similar "key" characters regardless of the actual botanical relationships involved. For example, since both Azalea (#158, in the Ericaceae, or Heath family) and the Geiger Tree (#429, in the Borgainaceae, or Borage family) are woody shrubs, have pink or red flowers, and share a few other characteristics, they "key out" together in R-1 although they are not botanically related. In some cases the two plants are keyed further, to their respective genera, but to save space, two, or even three, plants or plant groups may be indicated at the end of some keys, and, where a series of related plants of one plant family all fit one section of a key, only the family is listed along with the number of the first species in the family that has the characteristic indicated.

It will be much easier to key out a plant after you have looked at it closely and noted its main characteristics by comparison with the "key character summary code" inside the back cover.

Two of the most obvious characteristics of a plant are the color of its flowers and its habit, whether it is a woody tree or shrub, a vine, or one of the many herbaceous plants. For this reason, the following keys to the 500 plants pictured in this book are divided by flower color and plant habit into a series of 16 small keys that are much easier to work with than a single large key. For example, if you make a wrong choice on flower color (which is sometimes easy to do when a flower is pale pink, or pale blue and almost white, or between lavender and pink), it is easy to try the plant in two keys to find the correct match. Of course a few plants, such as ferns and pine trees, don't have flowers and some plants such as Australian Pine (*Casuarina*, #271, which is not a pine but a flowering plant) have very small flowers that are seldom noticed. These plants are included in a separate key, Key A.

To find which key you would use for a particular plant just refer to the table on the opposite page.

KEY A

(Non-flowering plants—ferns and gymnosperms—or angiosperms in which flowers are not produced or are not evident)

1. Plant a tree, shrub or vine	2.
2. Plant a vine	3.
3. Leaves simple, entire; reproduction by aerial tubers	*Dioscorea* (42)
3. Leaves compound, lobed, dentate; reproduction by spores	*Lygodium* (10)
2. Plant a tree or shrub	4.
4. Plant a shrub; lvs. pinnately compound	*Zamia* (11)
4. Plant a tree; lvs. or lflets. needle-like or scale-like	Gymnosperms (12) or *Casuarina* (271)
1. Plant herbaceous	5.
5. Plant terrestrial; reproduction by spores	Ferns (1)
5. Plant aquatic or epiphytic	6.
6. Epiphytes, growing on trees	7.
7. Leaves pinnately lobed, broad, green	*Polypodium* (9)
7. Leaves linear or quill-like, greyish or reddish	*Tillandsia* (85)
6. Aquatics, floating (plant body <3 cm wide or long)	8.
8. Plant body flat, lobed	*Azolla* (3)
8. Plant body rounded, entire	*Lemna* (76)

KEY G-I

(Flowers or inflorescence predominantly green or greenish white)

Plant a tree, shrub or vine

1. Plant a vine	2.
2. Leaves compound	3.
3. Leaves bipinnately compound	*Ampelopsis* (408)
3. Leaves palmately compound or trifoliate	4.
4. Leaves palmately compound, leaflets 5	*Parthenocissus* (410)
4. Leaves trifoliate, leaflets 3	5.
5. Flowers paniculate; berry white or yellow (poisonous!)	*Toxicodendron* (250)
5. Flowers umbellate; berry black	*Cissus* (409)
2. Leaves simple	6.
6. Leaves entire; plant often thorny	*Smilax* (41)
6. Leaves lobed or dentate; plant not thorny	7.
7. Leaves ovate, cut or deeply lobed, >3 dm long	*Monstera* (72)
7. Leaves cordate, dentate, <3 dm long	*Vitis* (411)
1. Plant a tree or shrub	8.
8. Leaves compound	*Ptelea* (242)

8. Leaves simple .... 9.
 9. Leaves opposite .... Batis (270)
 9. Leaves alternate .... 10.
  10. Leaves lobed or dentate .... 11.
   11. Leaves deeply palmately lobed or cut, margins serrate .... *Ricinus* (199)
   11. Leaves unlobed, margins dentate .... *Myrica* (255)
  10. Leaves entire .... 12.
   12. Leaves fragrant; infl. axillary or subterminal ..... *Persea* (119)
   12. Leaves not fragrant; infl. terminal .... *Conocarpus* (325)

KEY G-2

(Flowers or inflorescence predominantly green or greenish white)

Plant an aquatic or epiphyte

1. Plant an epiphyte or parasite on hardwood trees .... 2.
 2. Plant woody .... *Phoradendron* (228)
 2. Plant herbaceous .... *Epidendrum* (56)
1. Plant aquatic .... 3.
 3. Leaves, or entire plant, floating .... *Pistia* (75)
 3. Leaves, or plant, not floating but growing in water .... 4.
  4. Lvs. cordate, not grass-like; infl. a spadix with a spathe . *Peltandra* (74)
  4. Lvs. linear, grass-like; infl. not a spadix .... 5.
   5. Inflorescence of tight flower clusters .... *Sparganium* (77)
   5. Inflorescence an open panicle .... *Cladium* (91) or *Zizania* (104)

KEY G-3

(Flowers or inflorescence predominantly green or greenish white)

Plant a terrestrial herb

1. Inflorescence a spike or a spadix with spathe .... 2.
 2. Inflorescence a spadix with spathe .... *Arisaema* (70)
 2. Inflorescence a spike, solitary or in clusters .... 3.
  3. Lvs. inconspicuous, scale-like, opposite; plant succulent . *Salicornia* (260)
  3. Lvs. conspicuous, basal or alternate; plant not succulent 4.
   4. Leaves basal .... 5.
    5. Spikes several per scape; lvs. grass-like .... *Cyperus* (92)
    5. Spikes solitary; lvs. not grass-like .... *Plantago* (390)
   4. Leaves alternate .... 6.
    6. Infl. a raceme of spikelets; fruit spiny; lvs. linear, entire .... *Cenchrus* (100)
    6. Infl. a panicle of spikelets; fruit not spiny; lvs. lanceolate, dentate .... *Chenopodium* (259)
1. Inflorescence not a spike or spadix with spathe .... 7.
 7. Inflorescence an umbel .... *Asclepias* (364)
 7. Inflorescence not an umbel .... 8.
  8. Leaves basal, linear, grass-like .... Poaceae (99)
  8. Leaves cauline, opposite, not grass-like .... *Boehmeria* (192)

KEY W-1

(Flowers or inflorescence predominantly white or whitish)

Plant a tree, shrub or vine

Lead	Go to / Taxon
1. Plant a vine	2.
2. Leaves alternate	3.
3. Leaves compound	4.
4. Leaves bipinnate	*Cardiospermum* (251)
4. Leaves trifoliate or pinnate	5.
5. Leaves trifoliate (poison!)	*Toxicodendron* (250)
5. Leaves pinnate	*Galactia* (303)
3. Leaves simple	6.
6. Inflorescence a compact spike	*Peperomia* (121)
6. Inflorescence not a compact spike	7.
7. Flowers zygomorphic; parts in 3	*Vanilla* (69)
7. Flowers actinomorphic; parts in 5	Convolvulaceae (210)
2. Leaves opposite	8.
8. Flowers zygomorphic, or bilaterally symmetrical	*Lonicera* (423)
8. Flowers actinomorphic, or radially symmetrical	9.
9. Flowers solitary	*Rhabdadenia* (358)
9. Flowers in clusters	10.
10. Inflorescence knot-like	*Morinda* (351)
10. Inflorescence umbellate	*Echites* (356)
or	*Sarcostemma* (368)
1. Plant a tree or shrub	11.
11. Lvs. in terminal cluster; trunk unbranched	12.
12. Lvs. thin, simple, lobes not linear	*Carica* (176)
12. Lvs. leathery, palmately parted or pinnately compound, the lflts. or segments linear	Arecaceae (79)
11. Lvs. not all terminal; trunk branched	13.
13. Leaves linear, spine-tipped, in compact spiral around stem	*Yucca* (37)
13. Leaves not as above	14.
14. Leaves whorled	*Tetrazygia* (333)
14. Leaves opposite or alternate	15.
15. Leaves opposite	16.
16. Leaves pinnate	*Sambucus* (425)
16. Leaves simple	17.
17. Leaves lobed	*Hydrangea* (320)
17. Leaves unlobed	18.
18. Petals 4	19.
19. Inflorescence spherical	*Cephalanthus* (343)
19. Inflorescence not spherical	20.
20. Flowers tubular	Rubiaceae (345)
20. Flowers not tubular	21.
21. Leaves thin, deciduous	*Chionanthus* (229)
or	*Cornus* (412)
21. Leaves leathery, evergreen	22.
22. Shrub of tidal marshes	*Avicennia* (433)
22. Shrub of hammocks	*Myrcianthes* (328)
18. Petals 5	23.
23. Inflorescence racemose	*Brysonima* (233)
or	*Laguncularia* (326)

23. Inflorescence umbellate or cymose ...... 24.
24. Individual flowers <1 cm across ...... *Viburnum* (426)
24. Individual flowers >2 cm across ...... *Genipa* (346)
15. Leaves alternate .... 25.
25. Leaves compound .... 26.
26. Leaves palmate or trifoliate .... *Rubus* (282)
26. Leaves pinnate or bipinnate .... 27.
27. Leaves bipinnate .... *Aralia* (414)
27. Leaves pinnate .... 28.
28. Leaflets 9-23; rachis winged .... *Rhus* (248)
28. Leaflets 2-9; rachis not winged .... 29.
29. Leaflets 2; fruit a legume .... *Bauhinia* (290)
29. Leaflets 3-9; fruit not a legume .... 30.
30. Shrubs .... *Schinus* (249)
30. Trees .... 31.
31. Bark smooth, lustrous brown .... *Bursera* (246)
31. Bark not as above .... *Swietenia* (245)
or *Metopium* (247)
25. Leaves simple .... 32.
32. Individual fls. not evident, in compact heads . *Baccharis* (457)
32. Individual fls. evident, not in compact heads . 33.
33. Flowers solitary or appearing so .... 34.
34. Flowers >3 cm wide or long .... 35.
35. Petals 3 .... *Annona* (112)
or *Asimina* (113)
35. Petals 4 or more .... 36.
36. Stamens fused in tube around style .. *Hibiscus* (188)
36. Stamens separate .... 37.
37. Leaves serrate .... *Gordonia* (138)
37. Leaves entire .... *Magnolia* (109)
34. Flowers <3 cm wide or long .... 38.
38. Plant deciduous, thorny .... *Crataegus* (277)
38. Plant evergreen, thornless .... *Citrus* (241)
33. Flowers not solitary .... 39.
39. Leaves serrate .... 40.
40. Petals 4 or 7 .... *Ilex* (142)
40. Petals 5 .... 41.
41. Inflorescence racemose or spikate ... *Clethra* (143)
or *Itea* (321)
41. Inflorescence paniculate or umbellate 42.
42. Tree; inflorescence paniculate .... *Oxydendrum* (156)
42. Shrub; inflorescence umbellate ... *Ilex* (139)
or *Aronia* (275)
or *Prunus* (280)
39. Leaves entire .... 43.
43. Leaves 5 cm wide, leathery, round .... *Coccoloba* (265)
43. Leaves 5 cm wide or not leathery and round .... 44.
44. Corolla urceolate or funnelform .... 45.
45. Flowers perfect, not strongly scented .... Ericaceae (151)
45. Flowers unisexual, strongly scented *Croton* (196)

44. Corolla rotate or tubular .... 46.
46. Petals 4 .... *Ilex* (139) or *Capparis* (180)
46. Petals 5 or more .... 47.
47. Inflorescence spikate or racemose .... 48.
48. Inflorescence spikate; tree ... *Melaleuca* (327)
48. Inflorescence racemose; shrubs 49.
49. Inflorescence open, not compact .... *Befaria* (150)
49. Inflorescence compact ..... 50.
50. Shrub evergreen or semi-evergreen; swamps and wet areas .... Cyrillaceae (144)
50. Shrub deciduous; plant of drier uplands .... *Ceanothus* (193)
47. Inflorescence paniculate or umbellate .... 51.
51. Inflorescence umbellate ...... 52.
52. Leaves linear, fleshy, crowded, gray-tomentose .. *Tournefortia* (432)
52. Leaves round, not fleshy or crowded .... 53.
53. Leaves acute, deciduous; plant inland .... *Styrax* (165)
53. Leaves emarginate, evergreen; plant coastal .. *Chrysobalanus* (276)
51. Inflorescence paniculate ..... 54.
54. Panicles compact .... *Ardisia* (166)
54. Panicles open .... 55.
55. Shrub evergreen, <1 m tall .... *Licania* (279)
55. Shrub deciduous, >1 m tall .... *Solanum* (209)

KEY W-2

(Flowers or inflorescence predominantly white or whitish)

Plant an aquatic, epiphyte or parasite

1. Plant a leafless, non-green parasite .... 2.
2. Plant an orange vine on other plants .... *Cassytha* (117) or *Cuscuta* (211)
2. Plant white or yellow-brown, terrestrial .... *Monotropa* (155) or *Conopholus* (392)
1. Plant with leaves; epiphytic or aquatic .... 3.
3. Plant epiphytic on trees .... *Peperomia* (121)
3. Plant aquatic .... 4.
4. Leaves or entire plant floating or submersed .... 5.
5. Leaves linear, submersed .... *Egeria* (19)
5. Leaves cordate, floating .... *Nymphaea* (133) or *Nymphoides* (375)

4. Leaves or entire plant not floating or submersed	6.
6. Leaves basal	7.
7. Leaves slender, grass-like	*Dichromena* (93)
or	Eriocaulaceae (97)
7. Leaves not grass-like, blades wider	*Sagittaria* (17)
6. Leaves cauline	8.
8. Leaves whorled	*Alternanthera* (261)
8. Leaves alternate	*Saururus* (122)
or	*Nasturtium* (183)

KEY W-3

(Flowers or inflorescence predominantly white or whitish)

Plant a terrestrial herb

1. Leaves predominantly basal	2.
2. Flowers solitary or appearing so	3.
3. Flower parts 5; flowers zygomorphic	*Viola* (170)
3. Flower parts 3 or 6; flowers actinomorphic	4.
4. Flowers from a spathe	*Sisyrinchium* (47)
4. Flowers not from a spathe	*Hymenocallis* (27)
or	*Zephyranthes* (33)
2. Flowers not solitary	5.
5. Infl. flat-topped or umbellate; fls. >10 and <1 cm wide	6.
6. Flowers subtended by large, white leafy bracts	*Dichromena* (93)
6. Flowers not subtended by large, white leafy bracts	7.
7. Plant with red sap; infl. wooly	*Lachnanthes* (39)
7. Plant without red sap; infl. not wooly	8.
8. Leaves elliptic, tomentose beneath	*Eriogonum* (266)
8. Leaves round, or grass-like; glabrous	*Eriocaulon* (97)
or	*Hydrocotyle* (418)
5. Infl. not flat-topped or umbellate; or, if umbellate, fls. <10 and >1 cm wide	9.
9. Flower parts 5 or more	10.
10. Flowers in a compact, multi-flowered knot	*Eryngium* (417)
10. Flowers solitary or in a 2-3 flowered cyme	*Thalictrum* (126)
or	*Parnassia* (322)
9. Flower parts 3 or 6	11.
11. Flowers zygomorphic, or bilaterally symmetrical	12.
12. Inflorescence a raceme	*Calopogon* (49)
12. Inflorescence a spike	*Spiranthes* (66)
11. Flowers actinomorphic, or radially symmetrical	13.
13. Inflorescence not a raceme or panicle	Liliaceae (22)
13. Inflorescence a raceme or panicle	14.
14. Inflorescence a panicle; lf. margin filamentous	*Yucca* (38)
14. Inflorescence a raceme; lf. margin not filamentous	*Aletris* (20)
or	*Chamaelirium* (23)
or	*Zigadenus* (34)
1. Leaves cauline, opposite or alternate	15.
15. Leaves opposite	16.
16. Flowers solitary	17.

17. Petals 5 .... *Stellaria* (263)
or *Bacopa* (382)

17. Petals 4 .... 18.

18. Flowers tubular .... Rubiaceae (344)

18. Flowers rotate .... *Chamaesyce* (194)
or *Rhexia* (331)

16. Flowers not solitary .... 19.

19. Inflorescence a panicle (flowers minute) .... *Iresine* (262)

19. Inflorescence not a panicle .... 20.

20. Petals 4 .... *Richardia* (352)

20. Petals 5 .... 21.

21. Infl. a few-flowered cyme; fls. >2 cm wide .... *Sabatia* (372)

21. Infl. a multi-flowered head or umbel; fls. <1 cm wide .... 22.

22. Inflorescence a head or knot .... *Hyptis* (444)
or *Eupatorium* (471)

22. Inflorescence an umbel or a raceme .... 23.

23. Flowers actinomorphic, umbellate .... *Asclepias* (361)

23. Flowers zygomorphic, racemose .... *Dicerandra* (442)

15. Leaves alternate .... 24.

24. Infl. a head of tubular disc fls. and/or spatulate ray fls. .... 25.

25. Heads with disc flowers only .... 26.

26. Heads in dense terminal spikes; stem winged .... *Pterocaulon* (491)

26. Heads in corymbs or panicles; stem not winged .... *Gnaphalium* (478)
or *Palafoxia* (487)

25. Heads with both disc and ray flowers .... 27.

27. Leaves not lobed .... *Aster* (455)

27. Leaves lobed .... *Anthemis* (454)
or *Bidens* (461)

24. Infl. not a head of tubular disc and/or ray fls. .... 28.

28. Flowers zygomorphic, or bilaterally symmetrical .... 29.

29. Leaves compound .... 30.

30. Leaves palmately compound .... *Baptisia* (288)

30. Leaves pinnately compound .... *Dalea* (298)
or *Tephrosia* (312)

29. Leaves simple .... 31.

31. Petals 5 .... *Viola* (171)

31. Petals 3 .... 32.

32. Plant branched; inflorescence compact .... *Polygala* (234)

32. Plant not branched; inflorescence open .... 33.

33. Spike strongly spiraled .... *Spiranthes* (68)

33. Spike not spiraled .... *Habenaria* (58)
or *Platanthera* (62)

28. Flowers actinomorphic, or radially symmetrical .... 34.

34. Petals or flower parts 4 .... 35.

35. Leaves succulent or evergreen, shallowly lobed or cut .... *Cakile* (181)
or *Pachysandra* (200)

35. Leaves not succulent or evergreen, dentate or serrate .... *Lepidium* (182)
or *Gaura* (334)

34. Petals 5, or 10-12 .... 36.

36. Petals 10-12	*Anemone* (123)
36. Petals 5	37.
37. Flowers solitary, >5 cm across	38.
38. Petals separate, flat; lvs. spiny	*Argemone* (127)
38. Petals fused, tubular; lvs. not spiny	*Datura* (202)
37. Flowers not solitary, <5 cm across	39.
39. Inflorescence an umbel	40.
40. Umbel simple	*Solanum* (206)
40. Umbel compound	Apiaceae (415)
39. Inflorescence a raceme or spike	41.
41. Leaves lobed, with stringing hairs	*Cnidoscolus* (195)
41. Leaves unlobed, entire, without stinging hairs	*Polygonella* (267)
or	*Heliotropium* (430)

KEY Y-1

(Flowers or inflorescence predominantly yellow or orange)

Plant a tree, shrub or vine

1. Plant a vine	2.
2. Leaves trifoliate or pinnately compound	3.
3. Leaves trifoliate; stem herbaceous	*Vigna* (315)
3. Leaves pinnately compound; stem woody	Bignoniaceae (376)
2. Leaves simple	4.
4. Flowers zygomorphic	*Vanilla* (69)
4. Flowers actinomorphic	5.
5. Flowers tubular	*Gelsemium* (341)
or	*Urechites* (359)
5. Flowers rotate	*Passiflora* (174)
or	Cucurbitaceae (177)
1. Plant a tree or shrub	6.
6. Flowers appearing before or with developing leaves	7.
7. Infl. a catkin or drooping raceme	*Quercus* (272)
or	*Alnus* (274)
7. Infl. not a catkin, fls. in umbellate clusters or knots	Lauraceae (118)
or	*Symplocos* (164)
6. Flowers appearing after leaves are mature, or essentially so	8.
8. Leaves pinnately compound; flowers zygomorphic	9.
9. Leaves opposite	*Spathodea* (378)
9. Leaves alternate	10.
10. Flowers solitary	*Cassia* (292)
10. Flowers in a raceme or a knot-like head	*Acacia* (284)
or	*Caesalpinia* (291)
8. Leaves simple	11.
11. Leaves opposite	12.
12. Flowers tubular	*Angadenia* (353)
or	*Lantana* (436)
12. Flowers rotate	13.
13. Petals 4; plant with prop roots	*Rhizophora* (407)
13. Petals 5; plant without prop roots	*Hypericum* (146)
11. Leaves alternate	14.

14. Leaves serrate or lobed	15.
15. Leaves lobed	*Liriodendron* (108)
15. Leaves serrate	*Abutilon* (186)
14. Leaves entire	16.
16. Flowers solitary or appearing so	*Manilkara* (163)
or	*Suriana* (243)
16. Flowers in umbellate clusters or catkins	17.
17. Shrub or small tree; lvs. sticky; fruit winged	*Dodonaea* (252)
17. Large trees; lvs. not sticky; fruit an acorn	*Quercus* (273)

KEY Y-2

(Flowers or inflorescence predominantly yellow or orange)

Plant an aquatic, epiphyte or parasite

1. Plant an epiphyte on trees	2.
2. Flowers zygomorphic, elaborate and conspicuous; lvs. green	*Brassia* (48)
or	*Encyclia* (54)
or	*Epidendrum* (55)
2. Flowers actinomorphic, inconspicuous; lvs. grey-green	*Tillandsia* (87)
1. Plant aquatic	3.
3. Leaves floating	4.
4. Leaves slender, usually <3 cm wide	*Utricularia* (400)
4. Leaves wider, >3 cm wide, cordate	Nymphaeaceae (130)
3. Leaves not floating, but plant growing in water	5.
5. Lvs. slender or grass-like; infl. a compact, head-like spike	*Xyris* (88)
5. Lvs. not grass-like, blades wider; infl. an elongate fleshy spike	*Orontium* (73)

KEY Y-3

(Flowers or inflorescence predominantly yellow or orange)

Plant a terrestrial herb

1. Leaves absent or not obvious	2.
2. Stem fleshy, flattened, spiny; fls. regular	*Opuntia* (257)
2. Stem not fleshy, flattened or spiny; fls. zygomorphic	*Utricularia* (399)
1. Leaves present	3.
3. Inflorescence a knot-like head, or a head with rays	4.
4. Head knot-like, ray flowers reduced or absent	5.
5. Flowers all clearly zygomorphic	6.
6. Leaves opposite or basal	7.
7. Lvs. opposite; plant aromatic; stem square	*Leonotis* (445)
or	*Monarda* (446)
7. Lvs. basal; plant not aromatic; stem round	*Polygala* (236)
6. Leaves alternate	8.
8. Leaves elliptic; flowers orange	*Polygala* (238)
8. Leaves cordate; flowers yellow	*Rhynchosia* (307)
5. Flowers actinomorphic or appearing so	9.
9. Infl. axillary and terminal, asymmetrical or irregular; lvs. elliptic	*Waltheria* (185)
9. Infl. terminal, symmetrical; lvs. linear or lanceolate	10.

10. Leaves lanceolate; infl. elongate .... *Solidago* (493)
10. Leaves linear; infl. flat or rounded .... *Bigelowia* (462)
or *Euthamia* (474)
or *Flaveria* (475)
4. Head with obvious ray flowers .... 11.
11. Head of all petaloid ray flowers .... 12.
12. Leaves basal, not spiny .... *Hypochoeris* (482)
12. Leaves cauline, alternate, spiny .... *Sonchus* (495)
11. Head of both disc and ray flowers .... 13.
13. Leaves pinnate .... *Tagetes* (497)
13. Leaves not pinnate .... 14.
14. Leaves opposite .... *Tetragonotheca* (498)
or *Wedelia* (500)
14. Leaves alternate or basal .... 15.
15. Disc flowers maroon or brown .... *Coreopsis* (466)
or *Helianthus* (480)
or *Rudbeckia* (492)
15. Disc flowers yellow .... 16.
16. Rays lobed .... *Balduina* (459)
or *Berlandiera* (460)
or *Helenium* (479)
16. Rays entire .... 17.
17. Leaves succulent .... *Borrichia* (463)
17. Leaves not succulent . ... *Phoebanthus* (488)
or *Pityopsis* (489)
3. Inflorescence not a head .... 18.
18. Flowers zygomorphic .... 19.
19. Flowers solitary, or sometimes paired .... 20.
20. Flowers axilliary .... 21.
21. Leaves simple, opposite; fls. solitary .... *Gratiola* (385)
21. Leaves trifoliate, alternate; fls. paired .... *Stylosanthes* (310)
20. Flower scapose .... 22.
22. Leaves in flat basal rosette .... *Pinguicula* (396)
22. Leaves not in flat basal rosette .... 23.
23. Leaves finely dissected, inconspicuous .... *Utricularia* (399)
23. Leaves lanceolate, conspicuous .... *Viola* (169)
19. Flowers not solitary .... 24.
24. Leaves trifoliate or pinnately compound (fruit a legume) .... 25.
25. Leaves pinnate; flowers orange .... *Sesbania* (309)
25. Leaves trifoliate; flowers yellow .... *Baptisia* (289)
or *Crotalaria* (295)
24. Leaves simple .... 26.
26. Leaves cut, lobed or serrate .... Scrophulariaceae (381)
26. Leaves entire .... 27.
27. Flower parts 3 or 6; fruit not a legume ...... *Platanthera* (63)
or *Canna* (105)
27. Flower parts 5; fruit a legume .... *Crotalaria* (296)
or *Rhynchosia* (307)
18. Flowers actinomorphic or radially symmetrical .... 28.
28. Petals 3 or 6 .... 29.
29. Flowers solitary or appearing so .... 30.

30. Flowers nodding .... 31.
31. Flowers scapose; leaves mottled .... *Erythronium* (25)
31. Flowers axillary; leaves not mottled .... *Uvularia* (32)
30. Flower erect .... 32.
32. Flower >3 cm across, orange .... *Lilium* (29)
32. Flower <3 cm across; yellow .... *Hypoxis* (28)
29. Flowers not solitary .... 33.
33. Flowers in a slender raceme .... *Aletris* (21)
33. Flowers in a branched cluster .... *Lophiola* (40)
28. Petals 4, 5, or more .... 34.
34. Petals 4 .... 35.
35. Leaves pinnate, pinnately lobed, or spiny ...... *Argemone* (128) or *Raphanus* (184)
35. Leaves simple, entire .... 36.
36. Leaves opposite .... *Rhexia* (330)
36. Leaves alternate .... Onagraceae (336)
34. Petals 5 or more .... 37.
37. Petals more than 5 .... *Argemone* (128) or *Ludwigia* (335)
37. Petals 5 .... 38.
38. Leaves pinnate or trifoliate .... 39.
39. Leaves pinnate .... *Tribulus* (230)
39. Leaves trifoliate .... 40.
40. Flowers >5 mm wide; infl. open .... *Oxalis* (231) or *Duchesnea* (278)
40. Flowers <5 mm wide; infl. compact ..... *Thaspium* (422)
38. Leaves simple .... 41.
41. Leaves hollow, basal .... *Sarracenia* (134)
41. Leaves not hollow, cauline .... 42.
42. Flowers tubular with flared lip .... *Lithospermum* (431)
42. Flowers flat or shallowly funnelform .... 43.
43. Flowers umbellate or racemose, not solitary .... 44.
44. Inflorescence an umbel .... *Asclepias* (365)
44. Inflorescence a raceme .... *Piriqueta* (173)
43. Flowers solitary or appearing so ...... 45.
45. Lvs. opposite, scale-like; fls. <1 cm wide .... *Hypericum* (147)
45. Lvs. alternate, not scale like; fls. >1 cm wide .... 46.
46. Leaves clasping, weakly spiny .... *Argemone* (128)
46. Leaves neither clasping nor spiny . 47.
47. Flowers nodding; petals united . *Physalis* (204)
47. Flowers erect; petals separate .. *Ludwigia* (335)

KEY R-1

(Flowers or inflorescence predominantly pink or red)

Plant a tree, shrub or vine

1. Plant a vine .... 2.
2. Leaves opposite .... 3.

3\. Inflorescence a head; flowers pink .... *Mikania* (486)

3\. Inflorescence a few-flowered cyme; flowers red .... *Lonicera* (424)

2\. Leaves alternate .... 4.

4\. Stem thorny; inflorescence globose .... *Schrankia* (308)

4\. Stem not thorny; inflorescence not globose .... 5.

5\. Flowers zygomorphic; leaves compound .... *Abrus* (283) or *Galactia* (303)

5\. Flowers actinomorphic; leaves simple .... 6.

6\. Flowers in racemes, with 3 showy bracts .... *Antigonon* (264)

6\. Flowers not in racemes, without showy bracts ..... *Ipomoea* (214)

1\. Plant a tree or shrub .... 7.

7\. Flowers appearing with or before the young leaves .... 8.

8\. Flowers actinomorphic .... *Acer* (253)

8\. Flowers zygomorphic .... 9.

9\. Flowers <1 cm long or wide .... *Cercis* (294)

9\. Flowers >1 cm long or wide .... 10.

10\. Flowers solitary or appearing so, pink .... *Bauhinia* (290)

10\. Flowers racemose, red .... *Aesculus* (254)

7\. Flowers appearing after the leaves .... 11.

11\. Leaves opposite or whorled .... 12.

12\. Leaves whorled .... *Nerium* (357)

12\. Leaves opposite .... 13.

13\. Leaves palmately compound .... *Aesculus* (254)

13\. Leaves simple .... 14.

14\. Leaves 3-lobed .... *Acer* (253)

14\. Leaves not 3-lobed .... 15.

15\. Leaves serrate .... *Callicarpa* (434) or *Clerodendrum* (435)

15\. Leaves not serrate .... 16.

16\. Flowers tubular; all lvs. green .... *Hamelia* (348)

16\. Flowers not tubular; upper lvs. bright red .. *Poinsettia* (198)

11\. Leaves alternate .... 17.

17\. Leaves trifoliate, pinnate or bipinnate .... 18.

18\. Flowers actinomorphic; fruit not a legume .... Rosa (281)

18\. Flowers zygomorphic; fruit a legume .... 19.

19\. Leaves trifoliate .... *Erythrina* (301)

19\. Leaves pinnate (may be only 2 lflts.) or bipinnate 20.

20\. Leaflets 2-6 .... 21.

21\. Leaflets 2; flower solitary .... *Bauhinia* (290)

21\. Leafets 4-6; flowers in heads .... *Pitchecellobium* (306)

20\. Leaflets 10 or more .... 22.

22\. Flowers large, >4 cm long .... *Delonix* (299)

22\. Flowers small, <4 cm long .... *Albizia* (285) or *Tamarindus* (311)

17\. Leaves simple .... 23.

23\. Flowers zygomorhic; fruit a legume .... *Cercis* (294)

23\. Flowers actinomorphic; fruit not a legume .... 24.

24\. Flowers funnelform .... *Rhododendron* (157) or *Cordia* (429)

24\. Flowers rotate or cup shaped .... 25.

25\. Stamens united in column around style ...... Malvaceae (189)

25\. Stamens not united in column around style ... 26.

26. Fls. rotate; petals free; infl. a panicle ...... *Bixa* (172)
26. Fls. cup shaped, petals fused; infl. a raceme or fls. solitary .... *Kalmia* (152)
or *Lyonia* (153)

KEY R-2

(Flowers or inflorescence predominantly pink or red)

Plant an epiphyte or parasite

1. Epiphytes with red bracts or reddish, quill-like lvs. .... *Tillandsia* (84)
1. Terrestrial parasites or saprophytes without lvs. .... *Monotropa* (154)

KEY R-3

(Flowers or inflorescence predominantly pink or red)

Plant a terrestrial herb

1. Plant without lvs. or chlorophyll, entire plant pinkish or red . *Monotropa* (154)
1. Plant with lvs. and chlorophyll .... 2.
 2. Flowers in compact heads .... 3.
  3. Heads without rays .... 4.
   4. Leaves trifoliate or pinnate .... *Dalea* (297)
   or *Trifolium* (313)
   4. Leaves simple .... 5.
    5. Leaves spiny, lobed or serrate .... 6.
     6. Leaves opposite .... *Eupatorium* (472)
     6. Leaves alternate .... *Cirsium* (465)
     or *Pluchea* (490)
    5. Leaves entire .... 7.
     7. Leaves whorled .... *Polygala* (235)
     7. Leaves alternate .... *Marshallia* (485)
     or *Vernonia* (499)
  3. Heads with rays .... 8.
   8. Leaves dentate .... *Emilia* (469)
   8. Leaves entire .... *Coreopsis* (467)
   or *Gaillardia* (476)
   or *Lygodesmia* (484)
 2. Flowers not in compact heads .... 9.
  9. Flowers zygomorphic .... 10.
   10. Flower parts 3 or 6 .... 11.
    11. Flower solitary .... *Pogonia* (61)
    11. Flowers spikate or racemose .... 12.
     12. Flowers <2 cm long or wide; lvs. alternate ..... *Polygala* (237)
     12. Flowers >2 cm long or wide; lvs. basal .... *Calopogon* (50)
     or *Spiranthes* (65)
   10. Flower parts 4 or 5 .... 13.
    13. Leaves basal .... *Stenandrium* (406)
    13. Leaves cauline .... 14.
     14. Leaves trifoliate; fruit a legume .... *Desmodium* (300)
     or *Erythrina* (301)

14. Leaves simple; fruit not a legume	15.
15. Plant aromatic; stems square	*Dicerandra* (443)
or	*Salvia* (449)
15. Plant not aromatic; stems round	16.
16. Flowers crimson, tubular	*Lobelia* (224)
or	*Castilleja* (384)
or	*Dicliptera* (402)
16. Flowers pink, rotate or funnelform	17.
17. Plant succulent, fls. rotate	*Begonia* (179)
17. Plants not succulent; fls. funnelform	*Agalinis* (379)
9. Flowers actinomorphic	18.
18. Leaves opposite or whorled	19.
19. Petals 4	20.
20. Flowers rotate; plant not fleshy	*Rhexia* (329)
20. Flowers tubular; plant fleshy	*Kalanchoe* (317)
19. Petals 5 or more, or petal number not evident	21.
21. Inflorescence a panicle or raceme	22.
22. Inflorescence a panicle; petals free	*Sabatia* (371)
22. Inflorescence a raceme; petals fused	*Agalinis* (379)
21. Inflorescence not a panicle or raceme	23.
23. Flowers tubular	24.
24. Flowers in a spike	*Verbena* (440)
24. Flowers solitary or in umbellate clusters	25.
25. Flowers scarlet	*Spigelia* (342)
25. Flowers pink	*Catharanthus* (355)
23. Flowers rotate	26.
26. Flowers solitary	27.
27. Flowers large, >2 cm across	*Sabatia* (374)
27. Flowers small, <2 cm across	28.
28. Plant not succulent; leaves ovate	*Anagallis* (167)
28. Plant succulent; leaves linear	*Sesuvium* (256)
or	*Portulacca* (258)
26. Flowers clustered	29
29. Infl. an umbel; petals separate, reflexed	*Asclepias* (360)
29. Infl. a corymb; petals fused, not reflexed	*Phlox* (222)
18. Leaves alternate or basal	30.
30. Leaves predominantly basal	31.
31. Flower parts 3; lvs. linear	*Cuthbertia* (95)
31. Flower parts 5; lvs. not linear	32.
32. Flowers spurred; lvs. trifoliate	*Aquilegia* (124)
32. Flowers not spurred; lvs. simple and hollow or glandular	*Sarracenia* (135)
or	*Drosera* (323)
30. Leaves alternate	33.
33. Flowers tubular	*Ipomopsis* (221)
33. Flowers rotate	34.
34. Flowers small, < 1 cm across	35.
35. Flowers in a raceme	Polygonaceae (268)
35. Flowers solitary	*Geranium* (232)
34. Flowers larger, >1 cm across	36.
36. Petals 4	*Oenothera* (339)
36. Petals 5, or 10-12	*Anemone* (123)
or	*Callirhoe* (187)

KEY B-1

(Flowers or inflorescence predominantly blue, purple, lavender, maroon or brown)

Plant a tree, shrub or vine

1. Flowers maroon or brown .... 2.
  2. Plant a tree with slender, green, jointed, leaf-like branches . *Casuarina* (271)
  2. Plant a shrub or vine, branches not as above .... 3.
    3. Plant a vine .... 4.
      4. Leaves simple; flowers regular .... *Matelea* (367)
      4. Leaves trifoliate; flowers zygomorphic .... *Apios* (287)
    3. Plant a shrub .... 5.
      5. Leaves needle-like, <2 cm long .... *Ceratiola* (162)
      5. Leaves broad, >5 cm long .... *Calycanthus* (116) or *Illicium* (107)

1. Flowers blue, lavender or purple .... 6.
  6. Tree; leaves bipinnate .... *Melia* (244)
  6. Shrub or vine; leaves not bipinnate .... 7.
    7. Flowers zygomorphic, or bilaterally symmetrical .... 8.
      8. Plant a shrub .... 9.
        9. Leaves pinnate; plant not aromatic .... *Amorpha* (286)
        9. Leaves simple; plant aromatic .... *Conradina* (441)
      8. Plant a vine .... 10.
        10. Stem woody .... *Wisteria* (316)
        10. Stem herbaceous .... *Centrosema* (293) or *Vicia* (314)

    7. Flowers actinomorphic, or radially symmetrical .... 11.
      11. Leaves opposite .... 12.
        12. Vine; leaves pinnate or trifoliate .... *Clematis* (125)
        12. Shrub; leaves simple .... 13.
          13. Petals 4; inflorescence many flowered .... *Buddleja* (340)
          13. Petals 5; inflorescence few flowered .... *Stachytarpheta* (438)
      11. Leaves alternate .... 14.
        14. Flowers in compact heads .... *Garberia* (477)
        14. Flowers solitary or in open panicles .... 15.
          15. Plant a vine .... *Passiflora* (174)
          15. Plant a shrub .... *Lycium* (203) or *Solanum* (208)

KEY B-2

(Flowers or inflorescence predominantly blue, purple, lavender, maroon or brown)

Plant an aquatic, epiphyte or parasite

1. Flowers brown or maroon .... 2.
  2. Plant aquatic or epiphytic on trees .... 3.
    3. Plant epiphytic on trees; flowers zygomorphic .... *Campylo centrum* (51)
    3. Plant aquatic; flowers actinomorphic .... 4.
      4. Flowers in dense spikes, lvs. linear, not jointed .... *Typha* (78)
      4. Flowers in umbels; lvs. tubular, joined .... *Oxypolis* (420)

2. Plant terrestrial .... 5.
5. Plant unbranched .... *Corallorhiza* (53)
5. Plant branched .... *Epifagus* (393)
1. Flowers blue, purple or lavender .... 6.
6. Plant epiphytic on trees .... 7.
7. Fls. obscure, regular, spikate; lvs. quill-like .... *Tillandsia* (86)
7. Fls. evident, zygomorphic, paniculate; lvs. absent or not quill-like .... *Ionopsis* (60)
6. Plant aquatic or a terrestrial parasite .... 8.
8. Leaves absent or not evident .... 9.
9. Plant terrestrial, flower stalk stout .... *Orobanche* (394)
9. Plant aquatic, flower stalk slender .... *Utricularia* (401)
8. Leaves present and evident .... 10.
10. Leaves, or entire plant, floating .... *Eichhornia* (89)
or *Brasenia* (129)
10. Leaves erect, emergent; plant rooted in soil .... 11.
11. Inflorescence a compact, erect spike .... *Pontedera* (90)
11. Inflorescence an open, often nodding, panicle .... *Thalia* (106)

KEY B-3

(Flowers or inflorescence predominantly blue, purple, lavender, maroon or brown)

Plant a terrestrial herb

1. Flowers brown or maroon .... 2.
2. Leaves absent; stems not green .... 3.
3. Plant unbranched .... *Corallorhiza* (53)
3. Plant branched .... *Epifagus* (393)
2. Leaves present; stems green .... 4.
4. Leaves not basal .... *Trillium* (31)
or *Cleistes* (52)
4. Leaves basal or essentially so .... 5.
5. Flowers fleshy, beneath the aromatic lvs. .... *Hexastylis* (115)
5. Flowers membranous, well above the non-aromatic lvs. 6.
6. Flowers racemose; leaves elongate .... *Manfreda* (36)
6. Flowers in scapose, compact head; leaves wide .... *Helianthus* (481)
1. Flowers blue, purple or lavender .... 7.
7. Flower parts 3 or 6; fls. from a spathe .... Iridaceae (43)
or Commelinaceae (94)
7. Flower parts 4 or 5; fls. not from a spathe .... 8.
8. Flowers 2-lipped, or zygomorphic .... 9.
9. Plant without green leaves .... *Orobanche* (394)
9. Plant with green leaves .... 10.
10. Leaves basal .... 11.
11. Flowers numerous, spikate .... *Lupinus* (304)
or *Salvia* (448)
11. Flower solitary, scapose .... *Viola* (168)
or *Pinguicula* (395)
10. Leaves cauline, alternate or opposite .... 12.
12. Leaves alternate .... *Lobelia* (225)
or *Penstemon* (387)

12. Leaves opposite .... Acanthaceae (404)
or Lamiaceae (447)

8. Flowers regular or actinomorphic .... 13.
- 13. Leaves alternate or basal .... 14.
  - 14. Individual fls. small, indistinct, in compact heads . 15.
    - 15. Heads without rays, infl. spikate or paniculate .. *Carphephorus* (464) or *Liatris* (483)
    - 15. Heads with petal-like rays or large bracts, infl. not spikate .... 16.
      - 16. Inflorescence branched .... *Erigeron* (470)
      - 16. Inflorescence unbranched .... 17.
        - 17. Leaves cauline .... *Stokesia* (496)
        - 17. Leaves basal .... *Elephantopus* (468)
  - 14. Individual fls. distinct, not in compact heads ..... 18.
    - 18. Flowers axillary, solitary or paired .... 19.
      - 19. Corolla funnelform, >3 cm long or wide .... *Datura* (202)
      - 19. Corolla rotate, <2 cm long or wide .... *Triodanus* (226)
    - 18. Flowers neither solitary nor axillary .... 20.
      - 20. Plant spiny .... *Solanum* (207) or *Hydrolea* (428)
      - 20. Plant not spiny .... 21.
        - 21. Fls. >2.5 cm across; plant of swamps, wet ditches .... *Hydrolea* (428)
        - 21. Fls. <2.5 cm across; plant of dry pinelands or roadsides .... *Wahlenbergia* (227) or *Amsonia* (354)
- 13. Leaves opposite .... 22.
  - 22. Inflorescence spikate or a knot-like head .... 23.
    - 23. Spike open; fls. >8 mm long .... *Buchnera* (383)
    - 23. Spike compact or knot-like; fls. <5 mm long ... 24.
      - 24. Plant erect .... *Verbena* (439)
      - 24. Plant prostrate .... *Phyla* (437)
  - 22. Inflorescence not spikate or capitate .... 25.
    - 25. Flowers not axillary .... Gentianaceae (369)
    - 25. Flowers axillary .... *Dyschoriste* (403) or *Ruellia* (405)

Index

Items in this list are indexed by entry number. Names within parentheses are not main entries but are mentioned in the text of the entry shown.

Index

Index

Index

Index